AF839648

Maintaining Distribution System Residuals Through Booster Chlorination

The mission of the Awwa Research Foundation (AwwaRF) is to advance the science of water to improve the quality of life. Funded primarily through annual subscription payments from over 1,000 utilities, consulting firms, and manufacturers in North America and abroad, AwwaRF sponsors research on all aspects of drinking water, including supply and resources, treatment, monitoring and analysis, distribution, management, and health effects.

From its headquarters in Denver, Colorado, the AwwaRF staff directs and supports the efforts of over 500 volunteers, who are the heart of the research program. These volunteers, serving on various boards and committees, use their expertise to select and monitor research studies to benefit the entire drinking water community.

Research findings are disseminated through a number of technology transfer activities, including research reports, conferences, videotape summaries, and periodicals.

Maintaining Distribution System Residuals Through Booster Chlorination

Prepared by:
J.G. Uber, **D.L Boccelli**, **R.S. Summers**, and **M.E. Tryby**
Department of Civil & Environmental Engineering
University of Cincinnati
Cincinnati, OH 45221

Sponsored by:
Awwa Research Foundation
6666 West Quincy Avenue
Denver, CO 80235-3098

Published by the

Awwa Research Foundation
Advancing the Science of Water®

IWA Publishing

DISCLAIMER

This study was funded by the Awwa Research Foundation (AwwaRF). AwwaRF assumes no responsibility for the content of the research study reported in this publication or for the opinions or statements of fact expressed in the report. The mention of trade names for commercial products does not represent or imply the approval or endorsement of AwwaRF. This report is presented solely for informational purposes.

Published by IWA Publishing, Alliance House, 12 Caxton Street, London SW1H 0QS, UK

Telephone: +44 (0) 20 7654 5500; Fax: +44 (0) 20 7654 5555; Email: publications@iwap.co.uk;
Web: **www.iwapublishing.com**

AwwaRF report number 90927F. Originally published by AwwaRF for its subscribers in 2003.
IWA Publishing version published 2004.

ISBN 978-1-84339-859-2

Contents

List of Tables

List of Figures

Foreword

The Awwa Research Foundation is a nonprofit corporation that is dedicated to the implementation of a research effort to help utilities respond to regulatory requirements and traditional high-priority concerns of the industry. The research agenda is developed through a process of grass-roots consultation with members, utility subscribers, and working professionals. Under the umbrella of a Five-Year Plan, the Research Advisory Council prioritizes the suggested projects based upon current and future needs, applicability, and past work; the recommendations are forwarded to the Board of Trustees for final selection.

This publication is a result of one of those sponsored studies, and it is hoped that its findings will be applied in communities throughout the world. The following report serves not only as a means of communicating the results of the water industry's centralized research program but also as a tool to enlist the further support of the nonmember utilities and individuals.

Projects are managed closely from their inception to the final report by the foundation's staff and large cadre of volunteers who willingly contribute their time and expertise. The foundation serves a planning and management function and awards contracts to other institutions such as water utilities, universities, and engineering firms. The funding for this research effort comes primarily from the Subscription Program, through which water utilities subscribe to the research program and make an annual payment proportionate to the volume of water they deliver. The program offers a cost-effective and fair method for funding research in the public interest.

A broad spectrum of water supply issues is addressed by the foundation's research agenda: resources, treatment and operations, distribution and storage, water quality and analysis, toxicology, economics, and management. The ultimate purpose of the coordinated effort is to assist water suppliers to provide the highest possible quality of water economically and reliably. The true benefits are realized when the results are implemented at the utility level. The foundation's trustees are pleased to offer this publication as a contribution toward that end.

Edmund G. Archuleta, P.E.
Chair, Board of Trustees
Awwa Research Foundation

James F. Manwaring, P.E.
Executive Director
Awwa Research Foundation

Acknowledgements

The authors wish to sincerely thank the following persons and water treatment utilities for their full cooperation and participation in the research described in this report:

- Dr. Lewis A. Rossman, National Risk Management Research Laboratory, U.S. Environmental Protection Agency, Cincinnati, OH
- Ms. Laura Jacobson, Mr. Gregory Kodweis, Mr. Mao Fang, Las Vegas Valley Water District
- Mr. David Hartman and Mr. Jack DeMarco, Cincinnati Water Works, Cincinnati, OH
- Mr. Ken Jones, Utilities Department, City of Midland, Midland, Texas
- Mr. Mark Knudson, Bureau of Water Works, City of Portland, Portland, OR
- Mr. K. Kalyan Raman, Second Taxing District Water Department, South Norwalk, CT
- Mr. Carl Spangenberg and Mr. Kenneth Thompson, Irvine Ranch Water District, Irvine, CA

The authors also acknowledge the help and guidance provided by the members of the project advisory committee (PAC), including Paul Boulos, MWSoft, Inc.; Stuart Crymble, Severn Trent Water Limited; Peter Martin, East Bay Municipal Utility District; and Dolan McKnight, North Central Texas Water District. We are most grateful for technical and management guidance from project officer Michael Caprara, whose patience was tested severely during the completion of this work. Finally, the authors also appreciate the expertise provided by Margret Koechling in helping to design and conduct the chlorine disinfectant decay experiments, Mike Zierolf and Dr. Marios Polycarpou at the University of Cincinnati in helping on various modeling activities and continually bringing new and useful perspectives to discuss, and especially to Dr. Lewis Rossman of the U.S.E.P.A., who participated actively in project meetings and who's work on the EPANET Programmer's Toolkit was critical to developing efficient and extensible computer algorithms.

Chapter 1

Report Summary

The practice of using disinfectants to safeguard against the acute effects of waterborne disease are well established and regulated in the United States. New regulations designed to minimize the risks associated with chronic exposure to disinfection by-products (DBPs), however, are changing the current regulatory framework. Changes in disinfection practices meant to comply with these new regulations may inadvertently increase risks associated with microbial contamination.

Under this new regulatory framework, the operation of disinfection processes are over-constrained. Chlorine doses made within the water treatment plant (WTP) must simultaneously provide adequate microbial inactivation, provide residual maintenance at all points in the distribution system, and comply with DBP and taste and odor thresholds. If chlorine is only applied within the WTP, there is only one opportunity to simultaneously meet all treatment objectives. A booster chlorination strategy involves multiple coordinated doses applied throughout the distribution system. Such a strategy physically separates the doses meant for disinfection efficiency (microbial inactivation) requirements at the treatment plant, from those meant for residual maintenance in the distribution system. This physical separation improves a utility's operating flexibility for meeting disinfection efficiency, residual maintenance, and DBP requirements simultaneously

This report discusses at length the potential advantages associated with booster disinfection, the kinetics of chlorine decay and trihalomethane (THM) formation under rechlorination conditions, and the use of network hydraulic and water quality models to locate booster disinfection stations and determine their dose characteristics. Below a summary of the key findings is provided for each of these areas.

1.1 Quantifying Booster Disinfection Benefits

A conceptual analysis of the operating constraints that affect conventional chlorination emphasizes that, with the adoption of booster chlorination, achieving microbial inactivation can be viewed as a separate treatment issue from that of maintaining a detectable residual. Adoption of such a policy could allow dose reductions to be made at the WTP without compromising other treatment objectives. These dose reductions may reduce DBP yields at the WTP, and the disinfectant mass conserved can be more efficiently applied at points located in the distribution system (as a quantitative analysis demonstrates), by installing booster stations with this objective in mind.

An analysis of booster disinfection efficiency presents a conceptual model that helps to explain why booster chlorination may reduce disinfectant use, and why some operating policies are better than others. Two effects were identified which explain the reduction in disinfectant mass made possible by a booster disinfection strategy. By strategically distributing the chlorine dose using booster chlorination, a "physical distribution effect" can reduce disinfectant use by better matching of the dose to the unique ultimate residence time of a water parcel; rather than dose large flows near the treatment plant to levels that may only be required by small demands at the system periphery, booster disinfection reduces disinfectant use by lowering the dose for large plant flow and boosting the relatively smaller flows with long residence times. The second effect derives from the possible dependence of chlorine decay kinetics on chlorine concentration. For example, the typical first-order kinetic assumption implies that oxidizable organic matter is in excess. Under this assumption the chlorine decay rate is proportional to chlorine concentration, and any reduction in chlorine concentration under a booster scenario would imply a corresponding reduction in the total chlorine dose required for residual maintenance. This second effect is called the "chemical kinetic effect."

1.2 Disinfectant Decay and THM Formation

A laboratory procedure was proposed for analyzing chlorine decay under simulated booster conditions. This procedure may be used to evaluate the potential for reductions in chlorine use under booster chlorination, and more generally to understand chlorine decay kinetics under these conditions.

An initial set of experiments evaluated disinfectant decay under booster conditions, and found it to be dependent on both the magnitude and time of the boost dose. At a particular boost dose time ("boost time"), increasing the dose magnitude resulted in increased disinfectant decay. Increasing the boost time (for a particular boost dose), however, resulted in decreased disinfectant decay. The effect of boost time may be understood by examining the chlorine demand, which is a surrogate for the extent of reaction between disinfectant and the natural organic matter reactive component. Thus increasing the boost time will increase the chlorine demand, leading to less reactive material remaining at the time of the boost, and thus to decreased disinfectant decay compared to an earlier boost dose – even if the concentration after the doses are equal.

An additional set of experiments were performed to compare chlorine decay for a single initial dose with that for an initial dose plus a booster dose, where the total chlorine dose was held constant. The results showed the ability of booster chlorination to maintain an adequate residual with *reduced* disinfectant mass is dependent on the source water and chlorination conditions. The chlorine demand of the initial doses was shown to be consistent with a saturation model. Therefore, when the initial doses required to maintain an adequate residual were relatively high, the disinfectant decay of the initial doses (both conventional and boost) was independent of concentration, resulting in similar concentrations immediately after boosting and producing no bulk decay reduction under booster disinfection. For waters requiring relatively low initial doses to maintain adequate residuals, the disinfectant decay of the initial doses was dependent on concentration, resulting in less decay for the lower initial boost dose. In this latter case, a higher disinfectant concentration (relative to conventional disinfection) is obtained after application of the boost dose, allowing the boost dose to (potentially) be reduced. This increased disinfectant concentration, however, tends to disappear over time because the higher chlorine concentration after the boost dose, combined with a smaller chlorine demand at the time of the boost dose (i.e. more reactive component remaining), tend to increase disinfectant decay rates after the boost dose.

THM formation under booster conditions showed no long-term reduction for any set of results, indicating THM formation is not solely dependent on disinfectant concentration. When a single chlorine dose is split between an initial and a booster dose, the THM concentration tended to be reduced prior to application of the boost dose because of the smaller amount of disinfectant applied. After application of the boost dose, however, the THM concentration reached the level of an equivalent single dose experiment. Thus *prior to* the boost dose THM formation may be lower under booster conditions, providing a benefit to consumers in a distribution system between the source and a booster station. THM formation was found experimentally to be linearly dependent on chlorine demand under single and multiple dose conditions.

Preliminary studies with chloramine systems suggested booster disinfection will not allow a significant reduction in the amount of ammonia initially applied to the water. Booster disinfection did, however, appear to be very successful at producing, and maintaining, a low ammonia concentration by recombining with the ammonia formed(Valentine et al., 1997) after several days of chloramine decay.

A second-order model, including a reaction component that represents natural organic matter, was developed to describe chlorine decay and THM formation under booster conditions. This model performed better than a traditional first-order model when describing disinfectant decay under boosting conditions. While not an accurate mechanistic description of the process, the relative accuracy of the second-order model is the result of a crude, but effective, accounting for decreasing reactive component as chlorine decays over time.

1.3 Booster Disinfection System Design and Analysis

A method is developed to determine good *locations* and *schedules* for multiple booster doses, by a systematic evaluation of alternative system designs. The method relies on a calibrated hydraulic and water quality network model to simulate the effect of booster disinfection. Formal optimization methods are used to identify booster system designs that minimize the total mass of disinfectant applied to the system, while maintaining residuals at a set of monitoring locations. The key to the method is the use of linear superposition to describe the effect of multiple booster dosages; this means that the residual concentration at a monitoring location and time is proportional to the dose at each individual booster station. The design algorithms are contained in a software application for Booster Disinfection Design and Analysis (BDDA), which uses Epanet version 2 to simulate network hydraulics and water quality. A User's Guide to this software application is included.

Various BDDA analyses were illustrated using three distribution systems, to illustrate the design method and methods of analyzing booster station performance. Results suggest that booster disinfection may reduce the mass of disinfectant necessary to maintain residuals in the distribution system, when compared to conventional disinfection practices. The number and location of the booster stations, as well as how they regulate disinfectant dose, are important factors affecting system performance. Results indicate that booster systems can achieve a relatively low and uniform disinfectant residual distribution while offering improved residual maintenance, compared to conventional practice.

For one example network, the optimal booster dose schedules were obtained simultaneously with selection of optimal booster locations. Results suggest that significant disinfectant mass savings, compared to a conventional disinfection scenario (application only at the treatment plant), may be possible through use of coordinated booster stations. The addition of two booster stations reduced the total dosage from 69.7 kg/day to 42.0 kg/day, and adding two more booster stations (total of 4) reduced the dosage further to 29.0 kg/day, a reduction of more than 50% compared to the conventional case. Furthermore, the booster station designs effectively redistributed disinfectant mass from the treatment plant into the distribution system, leading to a more uniform, and lower, residual throughout the distribution system for a typical daily usage cycle. The average residual concentration (considering all consumer nodes and a typical daily water usage pattern) was reduced from 2.6 mg/L for the conventional case, to 1.52 mg/L using 2 booster stations, and 0.98 using 4 booster stations. The residual standard deviation was also reduced from 1.0 mg/L for the conventional case, to 0.53 mg/L using 2 booster stations, and 0.27 mg/L using 4 stations. These typical results show that booster disinfection has the potential to reduce aggregate exposure of the population to chlorine (and potentially its by-products), while simultaneously improving residual coverage in the periphery. A further benefit of booster chlorination that was not quantified is increased operational flexibility for maintaining chlorine residuals as network and usage characteristics change over time.

1.4 A Reader's Guide

This report includes both laboratory results and those that describe mathematical procedures and algorithms. As such it may not be easily accessible, in its entirety, to any one utility administrator, consulting engineer, or research scientist. Following is an attempt to guide the reader to appropriate sections, or to highlight prerequisite background knowledge.

Chapter 2 is mostly written for a general technical audience and discusses the motivation for booster chlorination and its advantages and limitations, although section 2.2.2 attempts to quantify the advantages of booster chlorination through a mathematical analysis for an example distribution system. Chapter 3 is also of a general nature, discussing the findings of a survey of current practice, and summarizing the relatively few written reports on booster chlorination case studies.

Chapter 4 describes laboratory chlorine decay and THM formation experiments conducted under booster conditions, as well as preliminary experiments for one chloraminated source. This material is written to an audience with a basic knowledge of chlorine decay and disinfection by-product formation kinetics (although this background is reviewed briefly in section 4.1). Section 4.7 is mathematically oriented, as it describes a model for chlorine decay and THM formation that is tested under booster chlorination conditions. Those not interested in such quantitative analysis can skip this section and focus on the data analysis.

Chapter 5 describes optimal booster system design procedures and algorithms, and necessarily assumes a reader who is knowledgable in applied mathematics and in water distribution network hydraulic and water quality modeling. Chapter 6 gives case study examples that apply these procedures, and can be read independently of chapter 5 if the reader is interested mainly in the capabilities of the design methods and how they can be applied. These two chapters taken together will be of interest to consulting engineers interested in new design approaches based on the joint use of network models and optimization software. A user's guide for the software developed is provided in Appendix A.

Chapter 2

Introduction to Booster Disinfection

This chapter presents an overview of booster disinfection and discusses some situations when utilities may consider it to be an effective strategy for residual maintenance. The final part of the chapter is a quantitative analysis of booster disinfection efficiency, which illustrates that disinfectant mass may be saved through two mechanisms: a physical distribution effect and a chemical kinetic effect. The physical distribution effect is due to redistributing the disinfectant mass throughout the network, in better proportion to changes in flow rates from water source to final consumer. The chemical kinetic effect is due to the rate of disinfectant decay being (possibly) a function of disinfectant concentration, and thus less decay taking place if disinfectant concentrations are reduced through coordinated boosting. Chlorine decay data are presented to support the notion that decay kinetics may not be first-order, and that non first-order kinetics would affect disinfectant mass savings due to booster disinfection. Chlorine decay and THM formation kinetics under booster disinfection are discussed in detail in chapter 4.

2.1 Background

2.1.1 Regulatory Setting

Disinfectants have been used in drinking water treatment since the late 1800's, when it was discovered that the risks associated with microbial contamination were greatly reduced with their application. The practice of using disinfectants to safeguard against the acute effects of waterborne diseases such as cholera is well established and regulated. Current treatment requirements, under the Surface Water Treatment Rule (SWTR), specify that disinfection is necessary for all treatment plants influenced by surface water sources(Agency, 1990), affecting approximately 10,000 plants in the U.S.(Auerbach, 1994; AWWA, 1992a). Typically, disinfectants are applied within the water treatment plant (WTP). Dosages are applied to satisfy disinfection requirements that, under the SWTR, are specified via *CT* values. Given reactor conditions such as temperature, pH, disinfectant concentration, and a statistical descriptor of residence time, CT values are an accepted method of ensuring that specified levels of disinfectant efficiency are achieved.

Maintenance of a detectable distribution system disinfectant residual has been adopted as part of the SWTR as a means of preventing microbial contamination in the distribution system, and is designated as the best available control technology for compliance with the Total Coliform Rule(Shaw and Regli, 1999). Doses which meet disinfection requirements are, if necessary, augmented to maintain residuals in the distribution system. New regulations, now in effect and being promulgated, will affect disinfection and residual maintenance. Stages 1 and 2 of the Disinfectants/Disinfection By-Product Rule are designed to protect against the health effects associated with exposure to disinfection by-products (DBPs). (DBPs are formed when disinfectants react with a matrix of dilute organic and inorganic DBP pre-cursors present in raw and treated water.) The frequency and severity of health effects associated with exposure to DBPs are still debated, although DBPs and disinfectants themselves are suspected carcinogens(Bull and Kopfler, 1991), and epidemiological and animal studies have suggested a link between DBPs and pregnancy loss(Waller et al., 1998; Bielmeier et al., 2001). Such evidence has led to a sharper focus on disinfection practices within the plant and distribution system, from both engineering and regulatory perspectives(Haas, 1999; Shaw and Regli, 1999). The new rules set increasingly stringent maximum contaminant levels (MCLs) for the trihalomethane (THM) and haloacetic acid (HAA) DBP categories, and for disinfectants themselves. This new regulatory framework presents a utility with a set of competing treatment objectives which require the control of risks associated with microbial contamination and DBPs. The adoption of less reactive disinfectants such as chloramines, or limiting residual maintenance dosages to control DBP formation, may increase microbial risk as utilities attempt to comply with these new regulations(Shaw and Regli, 1999).

2.1.2 The Value of Maintaining a Residual

Disinfectant residuals are thought to control biological regrowth, prevent reinoculation of indicator and pathogenic organisms from the biofilm, protect against contamination from sewage intrusion or other breaches in system integrity, and signal such breaches by their absence(Haas, 1999). Recently, however, the precise role of a disinfectant residual in meeting these objectives has been questioned.

Studies have been conducted to evaluate the efficacy of distribution residuals for biological regrowth control and reinoculation of indicator organisms from distribution system biofilms. One such study conducted by LeChevallier et al.(LeChevallier et al., 1987) evaluated the effectiveness of physical treatments and residual maintenance for their ability to prevent the presence of heterotrophic organisms and specifically coliform bacteria in distribution system pipes. Residual maintenance was evaluated by holding samples and controls for an additional contact time and by adding sufficient chlorine to the samples to substantially increase concentrations over in situ residual levels. The researchers reported no statistically significant decrease in coliform or heterotrophic bacteria due to this additional chlorine contact. Furthermore, the authors found that heterotrophic and coliform bacteria in the bulk fluid that were once associated with biofilms (released due to sloughing) had developed a resistance to disinfectant attack. This finding is corroborated by van der Kooij et al.(van der Kooij et al., 1999), who report that low disinfectant residual concentrations and contact times of several hours produce *CT* values which would be ineffective against suspended gram negative bacteria, viruses, and protozoans. The authors state that coliform bacteria are routinely observed in the presence of residual disinfectants, suggesting that the absence of indicator organisms in the presence a residual may lead to erroneous conclusions about the microbial safety of the water.

By means of pipeline sampling, the LeChevallier et al. study(LeChevallier et al., 1987) showed that heterotrophic and coliform bacteria were not uniformly distributed among potential distribution system sources. Pipe material and the presence of sediments were found to influence heterotrophic plate count (HPC) and coliform yields. While the physical treatment methods and disinfection conditions evaluated were ineffective at eliminating bacteria, the reported patterns of bacteria isolation can be interpreted as resulting from historic distribution system conditions - disinfectant residual levels, treatment techniques, and distribution system maintenance and operation. These results suggest that environmental factors such as habitat can affect the control of biological regrowth and pathogen propagation in the distribution system.

Treatment approaches where the biological stability of distributed water is the main objective as opposed to suppressing biological regrowth with residual disinfectants have been successful. In the Netherlands, groundwater and river bank filtered surface water sources low in biodegradable organic carbon (BDOC) are intrinsically biologically stable and distributed with a minimal or no residual disinfectant(van der Kooij et al., 1999). Good engineering practices designed to prevent breaches of system integrity are also used. These practices include construction using approved materials that minimize biological regrowth and corrosion, maintenance of consistently high pressures, maintaining a leakage rate $< 5\%$, strategic installation of back flow preventers and break tanks, supervisory control for the prevention of cross connections, and employing optimized microbiological and water quality monitoring schemes.

2.1.3 A Role for Booster Disinfection

Logic suggests that *integrated* engineering practices (such as those indicated above) would be most effective in managing factors in the distribution system that affect biological stability and system integrity; it would be unwise to rely solely on any one practice – including maintenance of a residual disinfectant – as a "final barrier", or to expect a disinfectant residual to correct otherwise poor management practices. Maintenance of a residual is, however, likely to be one important part of any integrated set of distribution system practices, although opposing forces (outlined above) are pushing utilities to do so through a wise and carefully controlled use of disinfectant. It may become increasingly difficult for utilities to manage disinfectant residuals exiting the treatment plant without also actively managing them in the distribution system. Booster disinfection represents a technological "middle ground" in the argument between those who claim no residual is needed, and those who claim it must be maintained, at any cost.

Utilities operate their disinfection processes while attempting to achieve multiple competing treatment objectives. A balance must be struck between disinfection efficiency requirements, residual maintenance, and disinfection by-product (DPB) formation control; the manner in which disinfectants are deployed influences the nature of this balance. Application of disinfectants within the treatment plant is, however, only one possible option; booster disinfection is a viable alternative. When practicing booster disinfection, disinfectant doses that satisfy CT requirements are made at the treatment plant and are augmented by booster additions applied in the distribution system, essentially distributing the points of disinfectant application, for residual maintenance purposes, outside of the treatment plant.

The booster disinfection strategy allows utilities to directly address the need for compromise among multiple treatment objectives. When chlorine is applied solely within the WTP, there is only one opportunity to simultaneously meet all treatment objectives. In contrast, a booster disinfection strategy employs multiple coordinated doses applied throughout the distribution system. Such a strategy physically separates the doses meant for microbial inactivation at the treatment plant, from those meant for residual maintenance in the distribution system. This physical separation improves a utility's flexibility when meeting microbial inactivation and residual maintenance requirements; a utility can maintain residuals at challenging locations in the distribution system without subjecting the entire population to elevated levels of disinfectant (and perhaps DBPs). A uniform and consistent disinfectant residual – resulting from a booster disinfection strategy – may help to manage biological regrowth, since intermittent or chronic residual loss may contribute to its occurrence. Reducing the temporal variability of disinfectant concentrations may also lower chlorine taste and odor complaints, since such complaints can be related to concentration fluctuations.

2.2 Simplified Analysis of Booster Disinfection

Despite the significant number of utilities that employ booster chlorination, little has been written discussing its potential benefits over conventional chlorination practice. For our purposes, the term "conventional" chlorination refers to the application of chlorine within the WTP for disinfection of the water supply and for residual maintenance. In contrast, booster chlorination — while still requiring chlorine application at the treatment plant for disinfection — relies on chlorine application within the distribution system to maintain adequate chlorine residuals there. Thus, booster chlorination is a method for residual maintenance rather than for finished water disinfection.

As described earlier, utilities operate their conventional disinfection processes while attempting to achieve four competing treatment objectives. Chlorine doses made within the WTP must simultaneously; 1) provide adequate microbial inactivation, 2) provide residual maintenance at all points in the distribution system, 3) comply with disinfection by-product (DBP) formation limits, and 4) comply with taste and odor thresholds. These four operating constraints can be reduced to three different conventional operating scenarios.

In this section we will discuss these three operating scenarios and how they affect residual maintenance in a conceptual distribution system. This conceptual system will be used to frame general questions about the potential benefits of using booster chlorination for the three operating cases, and how these benefits may be dependent upon system characteristics. One main benefit of booster chlorination is the ability to decouple disinfection and residual maintenance treatment objectives. As we will see over the course of the discussion, however, the potential effectiveness of booster chlorination for residual maintenance can not be completely separated from disinfection requirements.

2.2.1 Constraints on Disinfection Practice

Consider an operating scenario for the distribution system shown in Figure 2.1a and 2.1b, which practices conventional disinfection. The system includes a WTP and one finished water tank connected via a single inlet/outlet line. The tank floats on the system operating on a typical daily fill/drain cycle, and the diagrams in Figure 2.1a and b illustrate situations when the tank is filling and draining, respectively. Finished water is delivered to the first customer with a free available chlorine concentration labelled C_1, which is sufficient both for disinfection of the water supply and for residual maintenance in the distribution system. For purposes of discussion, the chlorine residual concentration should be greater than some value C_{min} at all points of use and at all times, based on a desire to suppress biological regrowth and pathogen entry due to breaches in system integrity. Similarly, the residual concentration at the first consumer (the maximum concentration experienced by consumers) should always be less than some value C_{max}. This maximum concentration reflects concern about taste and odor issues, and also about DBP formation, whose kinetics may depend on chlorine concentration (and hence on the dose, or the concentration at the first consumer). While these issues will typically control C_{max}, the proposed MCL of 4.0 mg/L represents an upper limit. Thus the residual concentration at the first consumer, and everywhere else in the distribution system, is practically limited by $C_{min} \leq C \leq C_{max}$.

The chlorine concentration at the first consumer, C_1, is determined from the above constraints on distribution system chlorine residual, as well as disinfection requirements. Regarding the residual constraints, let C_R be the minimum concentration required at the first consumer in order that $C \geq C_{min}$ everywhere and at all times. The concentration C_R is determined by the residual at a critical location and time in the distribution system, perhaps associated with a critical, or maximum, finished water residence time (this is only an approximation, since the pipe wall demand can combine with residence time to determine the true

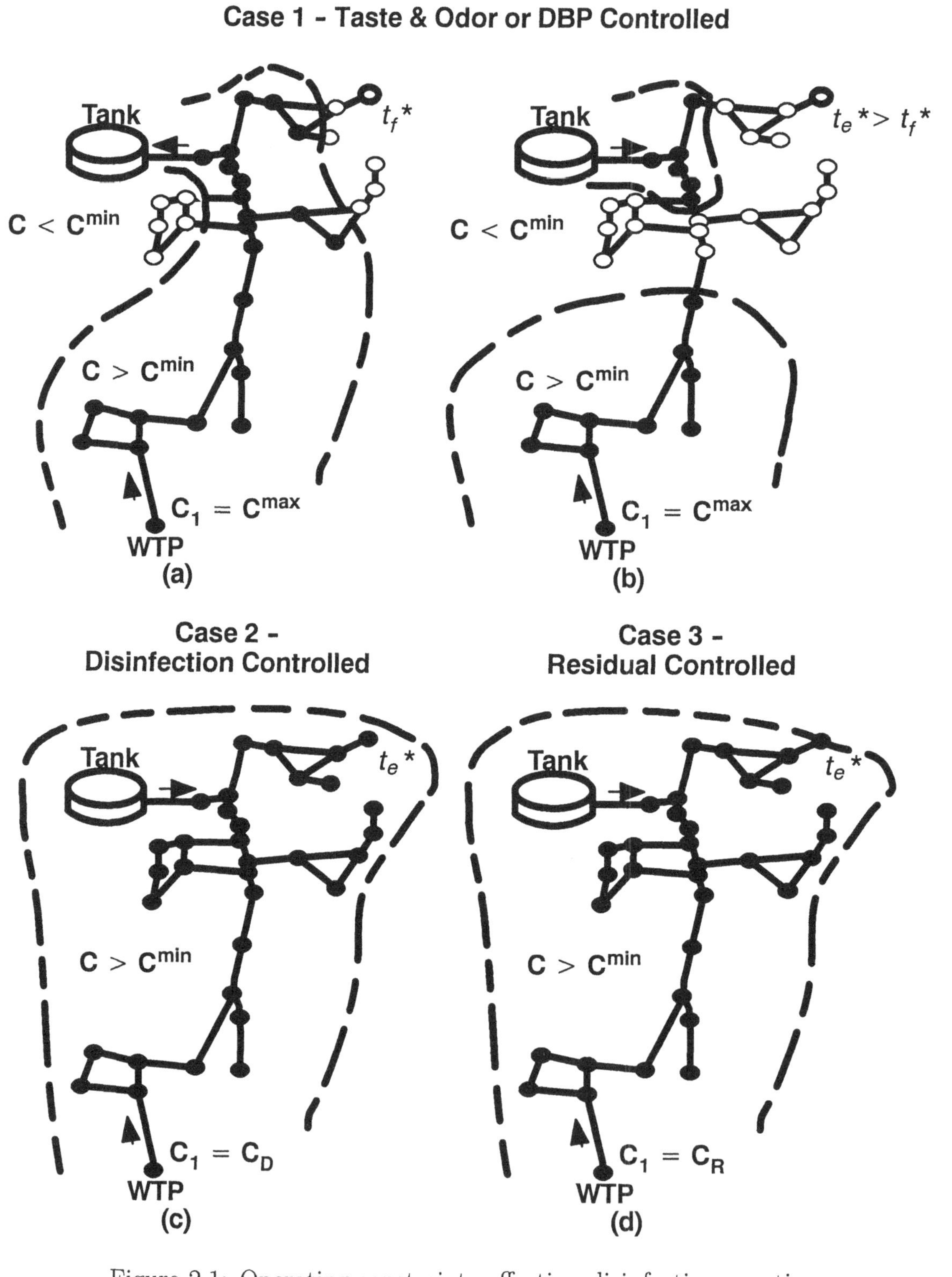

Figure 2.1: Operating constraints affecting disinfection practice

critical location). Regarding disinfection constraints, let C_D be the minimum concentration required at the first consumer for pathogen inactivation. Such a concentration would be determined by $C_D = CT_{min}/t_D^*$, where CT_{min} is the minimum concentration-time product, based on plant characteristics, and t_D^* is the characteristic disinfection contact time before the first consumer (e.g., the t_{10} value for a clearwell). Thus, three chlorine concentration constraints apply at the first consumer, characterized by the values of C_R, C_D, and C_{max}. The particular constraint that controls in a conventional disinfection scenario will affect the potential benefits from implementing booster chlorination for residual maintenance. Below we discuss the implications of booster chlorination under three possible scenarios: 1) taste and odor, or DBP controlled, 2) disinfection controlled, and 3) residual controlled.

Case 1 - Taste and Odor or DBP Controlled

If the concentration at the first consumer is controlled by taste and odor or DBP issues, then minimum distribution system residuals can not be maintained at all locations and at all times: $C_D < C_1 = C_{max} \leq C_R$. As illustrated in Figure 2.1a and b, while the finished water moves through the distribution system, filling the tank (Figure 2.1a), the chlorine residual decays due to reactions with natural organic matter and pipe wall materials. During the tank filling period the average travel time in the distribution system is relatively small, and thus much of the distribution system receives water with an adequate chlorine residual, $C > C_{min}$. Yet even during this period there may exist some areas of the distribution system that, due to long travel times or high pipe wall demand, receive water with inadequate residual, $C < C_{min}$; such areas, illustrated by white-colored nodes in Figure 2.1, are out of residual compliance. In the absence of pipe wall demand (which can be significant), the minimum residual location coincides with the longest average residence time, equal to t_f^* when the tank is filling. Figure 2.1a suggests that the node associated with the maximum average residence time t_f^* is located at the geographically furthest distance from the source, though that need not always be the case.

Distribution system areas that are below the minimum residual can not be brought into compliance — under conventional chlorination — without exceeding the upper limit C_{max} at the first consumer. The dashed lines in Figure 2.1a and b represent the residual compliance boundary, separating the distribution system consumption points into two sets: those that receive water with adequate chlorine residual, and those that do not. With hourly, daily, or seasonal changes in consumer demands, pump operation, or water quality parameters, the location of the residual compliance boundary will shift — as suggested in Figure 2.1b, which corresponds to periods when the tank is emptying. In this case the locations receiving water from storage will notice a reduced chlorine residual due to the additional tank residence time, as is suggested by the residual compliance boundary, which extends only a short distance from the tank. The maximum average residence time during the emptying period is longer than for the filling period, $t_e^* \geq t_f^*$, and thus the emptying period represents the critical period with respect to chlorine residual maintenance. In cases where the chlorine residual entering the tank is insufficient to maintain residual within the tank greater than C_{min}, all locations receiving stored water will be out of residual compliance during the tank draining period.

The above residual maintenance problems represent a straightforward case where booster chlorination can be an advantage over conventional chlorination: whenever maximum acceptable dosing at the plant is insufficient to maintain residuals at critical distribution system locations (associated with the longest travel times from the source, the pipes with the largest chlorine wall demands, or both). Booster dosing at strategic distribution system locations is then one means to maintain an adequate chlorine residual at consumption points. The addition of chlorine after storage in tanks would be common in this case, simply because the long residence times for stored water tend to deplete chlorine residual. Taste and odor or DBP controlled disinfection may become more common, given the factors pushing systems in this direction: prevalent use of finished water storage reservoirs for demand equalization; continued sprawl of urban areas and thus the distribution systems that deliver water; economies of scale associated with treatment facilities; geographic constraints that tend to place treatment facilities at the distribution system periphery (for surface water sources); and tighter regulatory constraints on DBP concentrations at points of consumption.

Cases 2 and 3 - Disinfection or Residual Controlled

If the concentration at the first consumer, C_1, is controlled by disinfection requirements, then distribution system residuals exceed the minimum required at all locations and at all times: $C_R < C_1 = C_D < C_{max}$. This case is illustrated in Figure 2.1c, which shows the critical residual case for the emptying tank. Similarly, if C_1 is residual controlled, then it is feasible to meet distribution system residual requirements under conventional chlorination, even though the dose will exceed that required for disinfection alone: $C_D < C_1 = C_R < C_{max}$. This case is illustrated in Figure 2.1d, where the residual compliance boundary passes through the critical location for residual maintenance.

At first glance, it might seem that a utility needn't consider booster chlorination if their conventional chlorination practice is disinfection or residual controlled; after all, both disinfection and residual requirements are being met, and C_1 is less than the maximum dictated by taste and odor or DBP concerns. Booster chlorination may, however, yield benefits that go beyond the ability to supply water with a minimum residual, and that relate more to how that residual is maintained throughout the distribution system and over the course of daily fluctuations in water use. Specifically, booster chlorination may allow a utility to meet residual targets with a reduction in total chlorine use that may represent a significant cost savings for some utilities.

Before discussing how booster chlorination can reduce chlorine use, we should briefly consider why such a reduction might be preferred. According to widely held disinfection and residual maintenance practice, beyond the need for disinfection of water supplies, a moderate residual should be kept in the distribution system to help control biological regrowth and as a guard against pathogen entry into the distribution system. Another view encourages a minimal and uniform residual and discourages excessively large chlorine additions from both a taste and odor and, potentially, a human health perspective (with this practice the population is exposed to lower concentrations of chlorine and, potentially, its reaction by-products). While such a view is not necessarily prevalent in the United States, it is more common in Europe, where chlorination for disinfection and residual maintenance are sometimes considered explicitly as two separate problems with different objectives. For example, the Paris treatment and distribution system employs chlorination for disinfection of their finished water, but dechlorinates after adequate disinfection is achieved. They have developed specialized telemetry and instrumentation sufficient to manage residuals as low as 0.1 mg/L throughout the distribution system, and employ booster chlorinators within the distribution system to do so (Nguyen and Montiel, 1994a).

Returning to Figure 2.1d and the case of residual controlled chlorination, the approach to residual maintenance discussed immediately above can be realized by installing booster chlorination stations within the distribution network. Once booster stations are installed, the minimum chlorine concentration at the first consumer, C_R, is reduced, since the critical residence time is shortened from that associated with the most distant location, t_e^*, to that associated with the most distant booster station. The concentration at the first consumer needn't be sufficient for the last consumer, and thus the chlorine doses for residual maintenance are metered out more uniformly throughout the distribution network, yielding a lower average chlorine concentration. There are limits to any reduction in chlorine concentration at the first consumer, since inevitably a residual controlled chlorination plan will become disinfection controlled when $C_R < C_D$. The implication is that, to deliver uniform and minimal chlorine residuals under booster disinfection, a utility should minimize C_D, either through plant changes that yield additional CT credits, or clearwell changes to increase the characteristic contact time t_D^*. Thus, booster chlorination to minimize and evenly distribute chlorine residual is not, in general, a separate issue from the design of plant and clearwell for disinfection (unless dechlorination is employed, as in the city of Paris case discussed above).

Decoupling Disinfection and Residual Maintenance

Booster disinfection allows dosages meant for residual maintenance to be shifted from the treatment plant to points located in the distribution system. Doses made within the treatment plant, however, must still achieve microbial inactivation. To the extent that dosages meant for these two purposes can be identified and separated, the adoption of booster chlorination allows these treatment objectives to be decoupled. Decoupling operations in this manner may provide utilities additional flexibility without compromising either treatment objective. Specifically, if a treatment plant is taste and odor, DBP, or residual controlled as discussed above, adoption of booster chlorination could allow dose reductions to be made at the WTP while still achieving microbial inactivation. The disinfectant mass conserved by these reductions can be more efficiently applied at points located in the distribution system by installing booster disinfection with this objective in mind.

2.2.2 Analysis of Booster Disinfection Efficiency

To this point we have described in general terms the potential for reduced disinfectant decay when using booster systems for residual maintenance. Characterizing these disinfectant reductions, however, is a difficult issue, due to complex distribution system hydraulics and chlorine decay kinetics. In this section we employ a mathematical analysis to quantify the disinfectant decay reductions and attribute them to either: 1) a physical distribution effect, or 2) a chemical kinetic effect. By mathematical analysis, these two effects are illustrated here using a simple branched network containing a single booster station.

Consider a distribution system consisting of a pipe header serving two different zones, such as that shown in Figure 2.2. Finished water is delivered at flow rate Q_0 (L^3/T) from the WTP to a junction with a travel time t_J (T). At the junction it is divided between two zones: flow rate Q_1 (L^3/T) is delivered to Zone 1, and flow rate $Q_2 = Q_0 - Q_1$ is delivered to Zone 2. Each zone has a single critical location with respect to chlorine residual, such that if residual targets are met at this critical location, they are met everywhere in the zone. The critical locations in Zones 1 and 2 are labelled points A and B, respectively. These might correspond to locations of maximum travel time within each zone, or perhaps to areas of high chlorine wall demand (or both). Because it is straightforward, and sufficient for the goals of this simple analysis, we assume that critical locations A and B are associated with maximum zone travel times t_A (T) and t_B (T), with $t_A \leq t_B$, and that wall demand needn't be considered. Thus, the total travel time to Zone 2 ($t_J + t_B$) is greater than that to Zone 1 ($t_J + t_A$) and so, from a system-wide perspective, point B is the critical residual location. The conventional dose at the WTP is assumed to be residual controlled (Case 3) so that a conventional chlorination scenario is feasible, as is a reduction of this conventional dose under a booster chlorination scenario.

The question asked here is: What is the most effective strategy for residual maintenance in both zones? This question will be explored below in some detail; an approximate answer can, however, be argued on intuitive grounds, and will serve to motivate the following analysis. Under a conventional scenario, the chlorine dose at the WTP clearwell is selected to satisfy minimum residual requirements at the critical location (point B, residual controlled). The chlorine residual at all other locations, including the critical location in Zone 1 (point A), will thus exceed the minimum requirement. Under a booster scenario, the chlorine dose locations and rates can be better matched to the distribution system residence times and user demands - in this case to the residence times and demands in each zone. For the distribution system in Figure 2.2, one logical location for a booster station is in the line feeding Zone 2, after the junction. The chlorine concentration at the WTP is now selected to satisfy minimum residual requirements at point A, which has a shorter residence time from the plant; thus this concentration is lower when compared to the conventional case. Although an additional

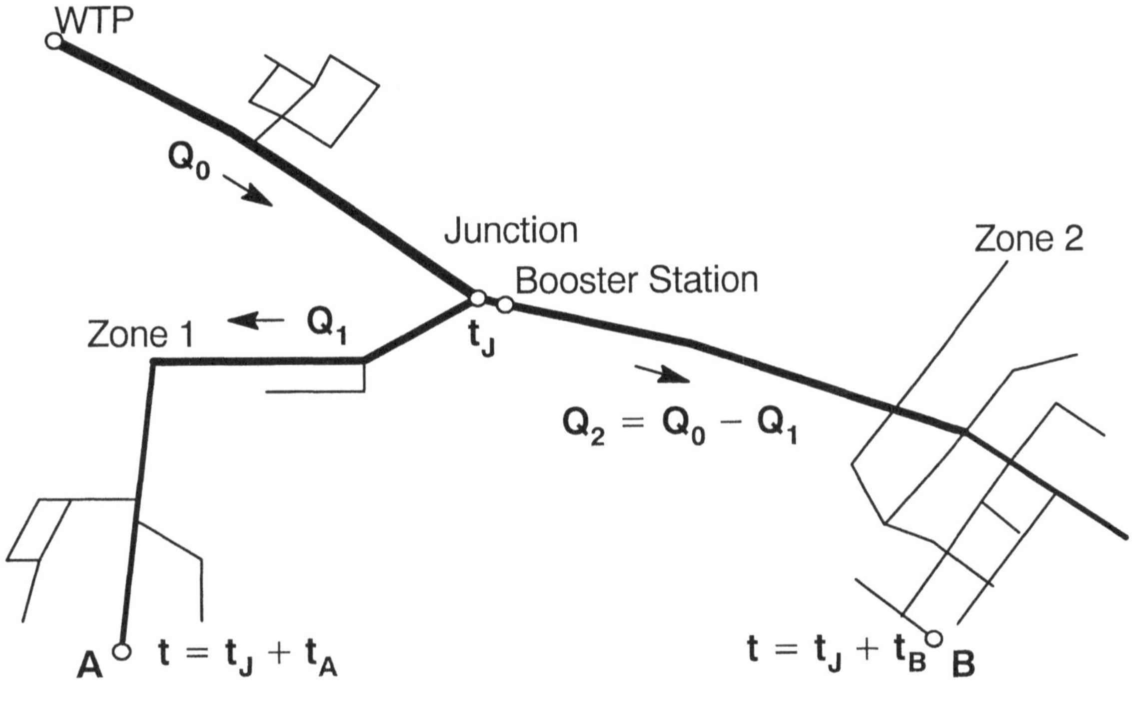

Figure 2.2: Booster disinfection in a hypothetical two-zone distribution system

dose is required to increase the residual concentration at the booster - sufficient to maintain a residual at point B — this dose rate increase must be smaller than the dose rate reduction at the WTP simply because the flow rate at the booster is smaller by the amount Q_1. In this manner, booster chlorination can lead to lower, more uniform residual concentrations within the distribution system and, at the same time, reduce the total dose rate that is needed for residual maintenance. This example also illustrates the importance of booster location; note that a booster station feeding Zone 1 (on the other side of the junction) would equate with the conventional strategy and thus accomplish nothing, since dosing at the WTP would remain controlled by point B.

Is booster chlorination more effective for residual maintenance compared to the conventional scenario? The answer is "yes," if the reduction in total chlorine dose results in significant cost savings (through lower chemical usage sufficient to offset booster station costs), or public health benefits (through reduced exposure to disinfection by-products or better control of biological regrowth). Unfortunately, the estimation of these benefits are complex and situation specific, and so do not allow a general answer to the above question. The potential cost and health benefits are, however, related approximately to the reduction in chlorine dose (cost saving being a more straightforward function of dose reduction than is health effect); thus, as a practical matter, we focus on the reduction in total chlorine dose as an approximate measure of booster chlorination effectiveness.

Using dose reduction as a measure of effectiveness, booster and conventional chlorination can be compared on a quantitative basis. An analysis for the distribution system in Figure 2.2 will suggest that the dose reduction attributed to booster chlorination is from two separable effects. The first has already been informally suggested: by strategically distributing the chlorine dose, booster chlorination can reduce chlorine use by better matching of the dose to the unique ultimate residence time of a water parcel. This first effect is called the "physical distribution effect." The second (and less obvious) effect derives from the dependence of chlorine decay kinetics on chlorine concentration. For example, the typical first-order kinetic assumption (for decay within the distribution system) implies that all substances reacting with chlorine are in such excess that their concentration is effectively constant. Under this assumption the chlorine decay rate is proportional to chlorine concentration, and any reduction in chlorine concentration under a booster scenario would lead to a reduction in the total chlorine dose required for residual maintenance. This second effect will be called the "chemical kinetic effect," and it is conceptually distinct from the physical distribution effect.

To conduct a quantitative comparison of the booster and conventional chlorination scenarios, and thus expose the physical distribution and chemical kinetics effects, the chlorine residual in the distribution system must be simulated under both scenarios. This task would be relatively simple if, for example, chlorine decay kinetics in the distribution system were first-order, and the same first-order decay coefficient could be used to simulate both conventional and booster scenarios. Unfortunately, this assumption is not always accurate, at least for chlorine decay due to bulk reactions (as opposed to wall reactions). Some experimental evidence shows that the chemical kinetic effect can be insignificant for bulk reactions, suggesting that the total chlorine demand may remain relatively constant irrespective of the rechlorination scenario (see chapter 4, also (Boccelli et al., 1997)). Thus disinfectant decay kinetics are not strictly first-order with respect to chlorine concentration. Contrasting first-order kinetics with constant chlorine demand (independent of the chlorination scenario) does, however, provide a simple means of bracketing the true behavior of bulk chlorine decay between two extremes. As shown below, these assumptions allow us to separate the physical distribution effect from the chemical kinetic effect.

Bulk Chlorine Decay Under Booster Conditions

To conduct a quantitative comparison of the booster and conventional chlorination scenarios, and thus expose the physical distribution and chemical kinetics effects, the chlorine residual in the distribution system must be simulated under both scenarios. This task would be relatively simple if, for example, chlorine decay kinetics in the distribution system were indeed first-order, and the same first-order decay coefficient could be used to simulate both conventional and booster scenarios. Unfortunately, the experimental data presented below will show that this assumption is not always accurate, at least for chlorine decay due to bulk reactions (as opposed to wall reactions). Indeed, the experimental evidence shows the chemical kinetic effect can be insignificant, suggesting that chlorine demand kinetics for bulk decay are independent of concentration. In reality we do not expect that disinfectant decay kinetics are either independent of or first-order with respect to chlorine concentration. These two extreme cases do, however, provide a simple means to bracket the true behavior of bulk chlorine decay, and thus constitute a conceptual model used to illustrate the limits of booster chlorination effectiveness.

Two sets of experiments were designed to simulate rechlorination conditions like those for the example distribution system in Figure 2.2. Water for the experiments was taken from Harsha Lake (located in southwestern Ohio), and was either pilot or bench scale alum coagulated prior to filtration with a 0.45 μm membrane in the laboratory. Experimental materials and methods are discussed in chapter 4.

Experiment 1 was designed to compare the kinetics of chlorine decay under "equivalent" conventional and booster conditions, in the sense that the total mass of chlorine added was identical. The conventional scenario simulated the single application of disinfectant at the clearwell of a treatment plant, while the booster scenario simulated application of the same chlorine mass, but divided it between the treatment plant and the booster station. Chlorine decay results are shown in Figure 2.3 for a 1.5:1 $Cl_2 : DOC$ (concentration) ratio. Figure 2.3a compares conventional with booster conditions where the treatment plant dose is 75% of conventional and the remaining 25% is applied at the booster station, while Figure 2.3b is similar except the booster conditions split the doses equally between the treatment plant and the booster station. In both cases, the booster dose was applied after the chlorine concentration had decayed to approximately 0.2 mg/L, simulating a practical situation where low residuals are replenished to meet target concentrations in more remote regions.

The results in Figure 2.3 suggest the absence of any appreciable chemical kinetic effect that would attenuate bulk chlorine decay under booster conditions. Indeed, the chlorine decay is insensitive to chlorine concentration in this case, as the residual concentration after the booster dose is close to that for the conventional dose. In contrast, a first-order model for bulk chlorine decay kinetics — using the same rate constant for the conventional and booster scenarios — would predict a higher chlorine concentrations after the booster dose compared to the conventional single dose. Thus, a first-order model may not always be appropriate for estimating the benefits of booster chlorination.

This latter point concerning the first-order model is worth emphasizing. A natural approach to estimating the practical benefits of booster chlorination is through the use of distribution system water quality models, which typically assume first-order kinetics for bulk chlorine decay. Thus the same potential modelling errors alluded to above (for the Figure 2.3 results) would apply if a calibrated distribution system model were used to simulate booster conditions by simply changing the chlorine dose pattern. The first-order assumption is still valuable as an approximate basis for design, but the results should be interpreted with care. Furthermore, the analysis and experimental results presented here do not consider wall demands, which can be an important process for chlorine decay in distribution

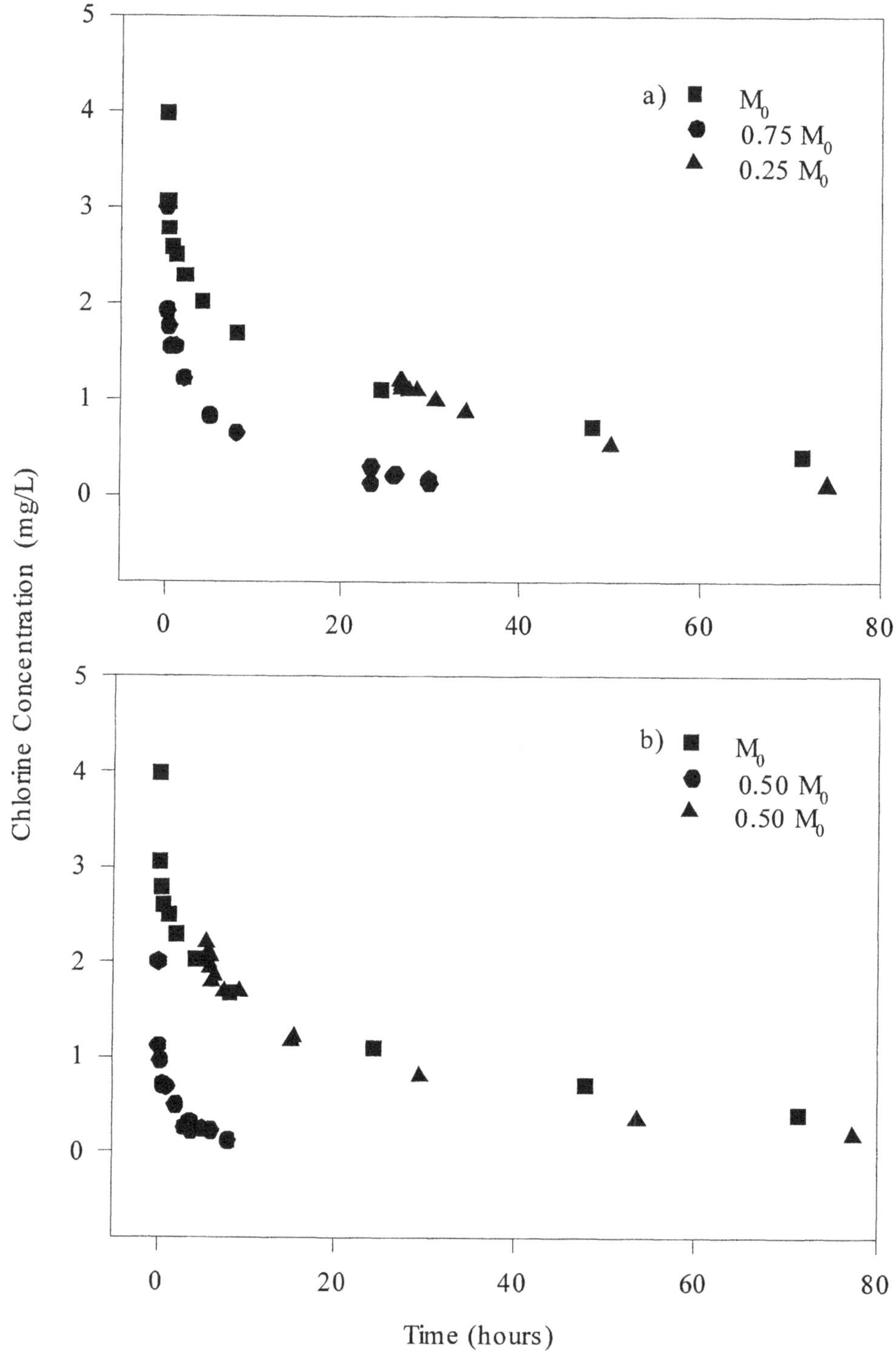

Figure 2.3: Chlorine decay under equivalent mass boosting and non-boosting conditions. (a) 25% Boost dose (b) 50% Boost dose

systems, and can be modelled as a first-order process dependent on hydraulic and pipe characteristics(Rossman et al., 1994). Thus these results do not speak to the adequacy of such first-order models for estimating wall demand under booster conditions.

The Figure 2.3 results prompted a more systematic exploration of chlorine decay kinetics under rechlorination conditions. A second experiment explored the effects of rechlorination time and mass concentration on chlorine decay kinetics. Waters that were initially chlorinated at two different $Cl_2 : DOC$ ratios (1.5:1 and 2.5:1) were rechlorinated at one of three different times: 24 hours after the initial dose, 48 hours after the initial dose, and after the initial dose decayed to approximately 0.2 mg/L (168 hours after the initial dose). At each of these times the rechlorination dose was calculated to result in one of three concentrations after boosting: the initial dose concentration, the concentration 8 hours after the initial dose, and the average of these two concentrations. While the booster conditions simulated in Experiment 2 are no longer compared to an equivalent conventional scenario, these results can aid in understanding chlorine decay kinetics under boosting conditions.

Experiment 2 results are shown in Figure 2.4, for the 1.5:1 $Cl_2 : DOC$ ratio (the 2.5:1 ratio data show similar trends and thus are not shown), as well as estimated first-order model solutions, $C(t) = C_0 e^{-k_b t}$ (solid lines), corresponding to estimated values of the first-order bulk decay rate coefficient k_b (hr^{-1}) computed by least-squares regression. Each initial and rechlorination dose data set is modeled using a unique rate coefficient (shown in Figure 2.4), so the various model solutions illustrated do not constitute a single model of chlorine decay under booster conditions. Also, the initial dose models ignore the first 4 hours of chlorine decay data, corresponding to rapid reactions that would occur within the clearwell.

The first-order models shown in Figure 2.4 (solid lines) are adequate representations of experimental data. Note, however, that the estimated bulk decay coefficients vary significantly from the initial dose to the various booster dose models, and also among the different booster dose time and mass concentration models. Furthermore, these variations in k_b are systematic: decreasing with increasing rechlorination time (for the same booster dose concentration), or with increasing rechlorination dose concentration (for the same booster dose time). These systematic variations are consistent with a conceptual model of chlorine decay that includes reactions with oxidizable organic compounds; once oxidized these organic compounds no longer enter into such reactions, affecting the observed decay kinetics. A first-order model in chlorine concentration cannot account for such influences, and the systematic variation in k_b is an artifact of using such a simple approach.

If the first-order initial dose model ($k_b = 0.0178\ hr^{-1}$) is used to represent data after rechlorination, the bulk chlorine decay would be overestimated by a significant margin (dashed lines in Figure 2.4). It appears that, in order to accurately represent bulk chlorine decay under booster conditions using a first-order model, one may be forced to estimate unique decay coefficients for each unique rechlorination dose and chlorination history.

Bulk chlorine decay kinetics appear to be more complicated than what is represented by typical first-order models. This additional complexity is perhaps due to the changing

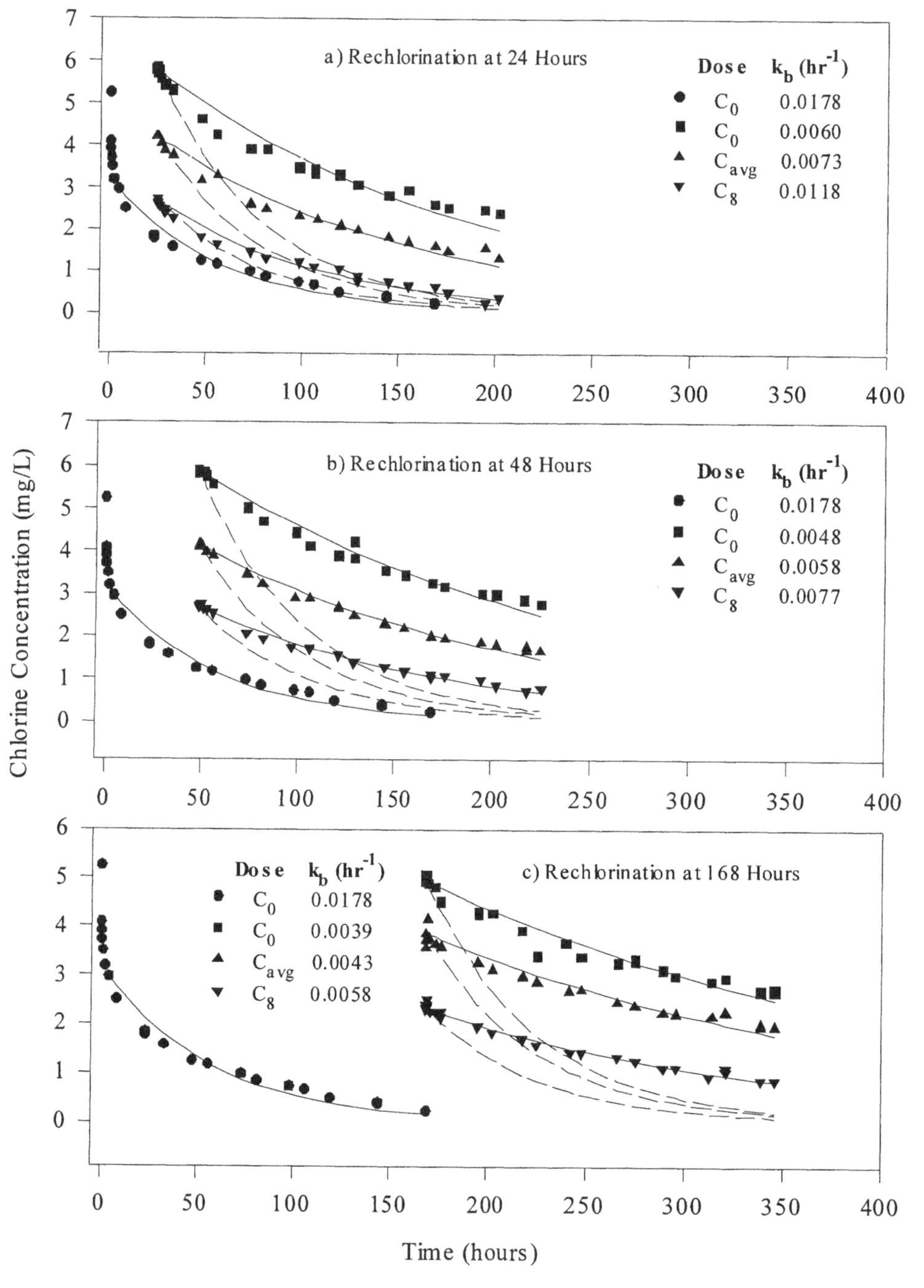

Figure 2.4: Chlorine decay kinetics for single dose and subsequent booster doses at three rechlorination times

concentrations of oxidizable organic matter that are ignored by such models, and that are in fact assumed to be in excess. Although the experimental results presented here are limited to one water, the same trends are duplicated in rechlorination experiments on three other waters also investigated by the authors (see Chapter 4. In any case, the purpose of exhibiting these results here is not to propose a more accurate model of bulk chlorine decay kinetics, but rather to illustrate some limitations that should be placed on the first-order model. These limitations will allow the first-order model to be viewed as one extreme model of behavior, and used as a limiting case for comparing the effectiveness of booster and conventional chlorination scenarios.

The comparison of conventional and booster chlorination for the distribution system in Figure 2.2 will require estimation of the chlorine demand in the bulk water under both scenarios. The relevant question can be stated: Given the travel times from the WTP to the junction, and from the junction to points A and B, how is the chlorine demand affected by differences in chlorine dosing between the conventional and booster scenarios? One view of the above experimental data that is relevant to this question is a plot of chlorine demand at some time t (the difference between initial chlorine concentration and that remaining at time t) versus $Cl_2 : DOC$ ratio (or, equivalently, versus initial chlorine concentration). Such a plot is shown for Harsha Lake water in Figure 2.5. The chlorine demand was calculated at 24, 48, 72, and 120 hours after the initial dose, and for five initial doses ranging from 3.5 to 10.2 mg/L (1:1 to 3:1 $Cl_2 : DOC$). The chlorine demand can never fall to the left of line labelled $t = \infty$, representing the condition where chlorine demand equals the dose.

The chlorine demand *at a specific time* depends strongly on the dose applied at lower initial doses, yet is independent of the dose applied at higher doses. Two different conditions govern this extreme behavior at the low and high ends of the dose spectrum, consistent with the conceptual model of chlorine-organic matter reactions used to explain observed deficiencies of the first-order model. At low initial doses, oxidizable natural organic matter (NOM) is in excess and chlorine concentration can be assumed to be limiting, yielding behavior that is approximately first-order with respect to chlorine concentration. At high initial doses, chlorine is in excess and oxidizable NOM can be assumed to be limiting, leading to chlorine demand kinetics that are independent of chlorine concentration (or dose). In this latter case the chlorine demand would be constant and independent of chlorine dose, and thus a decrease in dose would not yield any benefit in terms of a decrease in chlorine demand.

One can interpret the chlorine demand data using a first-order model, as follows. An initial rapid chlorine demand, CD_I (mg/L), is assumed to occur during the time interval $0 \leq t \leq t_I$, and this demand is assumed independent of the initial dose concentration, C_0 (mg/L). Assuming first-order chlorine decay kinetics, the chlorine concentration for all times $t > t_I$ is then $C(t > t_I) = (C_0 - CD_I)e^{-k_b(t-t_I)}$, and thus the total chlorine demand at time $t > t_I$, including the constant initial demand, is given by $CD(t > t_I) = C_0[1 - e^{-k_b(t-t_I)}] + CD_I e^{-k_b(t-t_I)}$. As illustrated in Figure 2.5, this first-order model assumes chlorine demand to be linear with respect to chlorine dose; the slope of the line, $[1 - e^{-k_b(t-t_I)}]$, equals 0 when $t = t_I$ (the chlorine demand equals the constant initial demand), and approaches 1

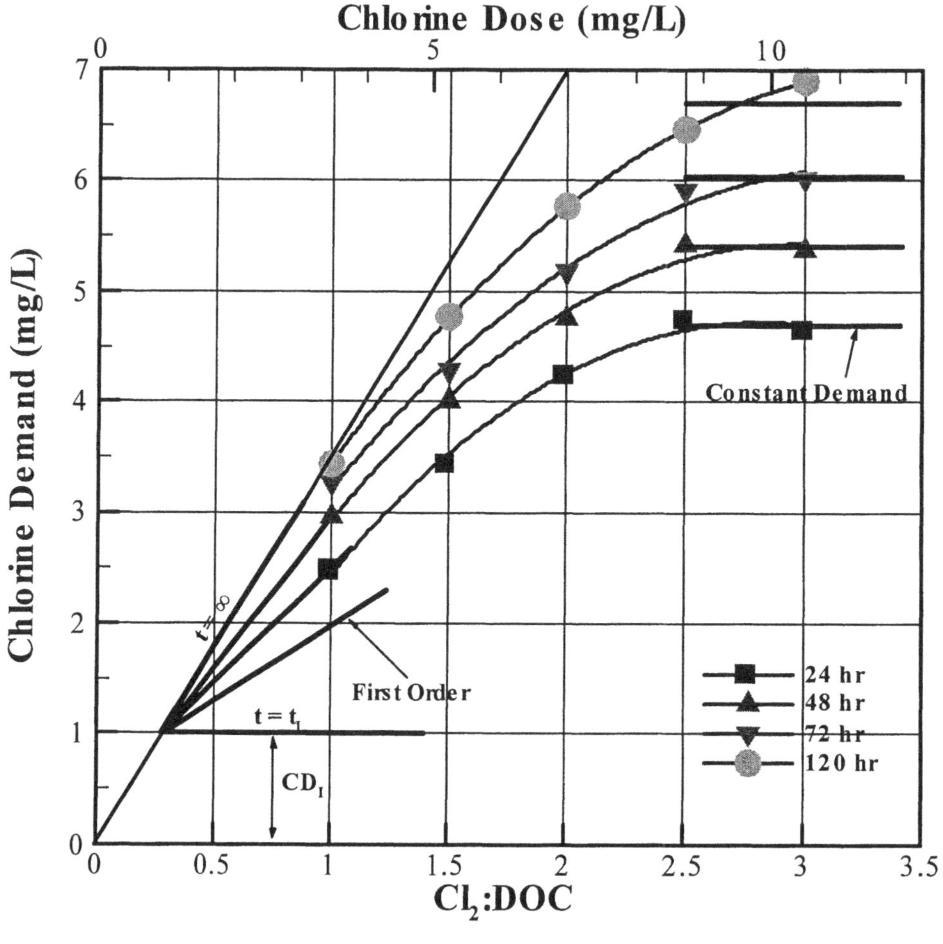

Figure 2.5: Chlorine demand versus initial dose for different measurement times

exponentially as $t \to \infty$ (at some sufficiently large time, all chlorine is demanded, no matter what the dose). Figure 2.5 suggests that the first-order model can be a good representation of the data for low chlorine doses, but that as dose increases so does the deviation from model behavior. Indeed, for sufficiently high doses — the opposite extreme case — the chlorine demand appears to become constant with respect to chlorine concentration.

The results in Figure 2.5 illustrate two extreme models that, it can be argued, bracket the range of observed chlorine decay behavior in a way that has meaning for booster chlorination efficiency. Each curve in the figure can be thought to correspond to a residence time (or, a location) in the distribution system of Figure 2.2 - either from the WTP to the junction, or from the WTP to points A and B. The addition of a booster chlorine dose at the junction allows a reduction in the initial dose at the WTP (compared to the conventional scenario), which corresponds to moving along one of the lines in Figure 2.5. The extreme models of chlorine decay kinetics correspond to assuming one of two cases: 1) linear change in chlorine demand from such a reduction (first-order, low $Cl_2 : DOC$); or 2) no change in chlorine demand from a reduction in initial dose (constant demand, high $Cl_2 : DOC$).

Quantitative Analysis

The above observations concerning chlorine decay kinetics can be used to illustrate the advantages of booster chlorination for the distribution system in Figure 2.2. The dose reductions possible with booster chlorination are a function of the *physical distribution* and *chemical kinetic* effects. The following analysis will separate these two effects, thereby illustrating the range of chlorine dose reductions likely to occur for the booster chlorination scenario analyzed here.

Booster efficiency is defined as:

$$\epsilon = \frac{M_0 - (M_1 + M_2)}{M_0} \tag{2.1}$$

where M_0 (M/T) is the chlorine mass rate required at the WTP to maintain an adequate residual at point B (conventional case), minus the mass rate consumed by any initial demand assummed to occurr within the WTP; similarly, M_1 (M/T) is the mass rate required at the WTP to maintain an adequate residual at point A (boosting case), and M_2 (M/T) is the supplemental mass rate applied at the booster station to maintain an adequate residual at point B. Thus 2.1 is simply the percent savings in total chlorine dose rate attributable to booster dosing, over and above that required to satisfy initial demands and assuming that disinfection requirements are met.

The mass rate M_0 is computed by assuming the chlorine decay kinetics for a particular conventional scenario are represented adequately by a first-order assumption. (Note: this assumption does not accept the first-order model as an adequate representation of chlorine decay in general. Rather, it only assumes that a conventional dose scenario can be represented by a first-order model and a single rate coefficient k_b (T^{-1}), in the same way that chlorine decay data is frequently represented using a first-order model in practice.) In this case, if the conventional WTP dose satisfies C_{min} (M/L^3) at point B, then $C_{min} = (M_0/Q_0)e^{-k_b(t_J+t_B)}$, or

$$M_0 = C_{min}Q_0e^{k_b(t_J+t_B)} \tag{2.2}$$

The computation of dose rates for the booster scenario, M_1 and M_2, depends on which of two extreme models of chlorine decay kinetics is adopted: constant or first-order. The rate of chlorine decay, for the first-order assumption, is a function of the decay rate coefficient and chlorine concentration; thus, the chlorine demand in each zone is solely a function of the rate of chlorine decay and the flow rate into that zone. Specifically, the first-order model yields the following dose rates for the booster scenario, assuming that the minimum chlorine residual is exactly satisfied at points A and B:

$$M_1 = C_{min}Q_0e^{k_b(t_J+t_A)}, \tag{2.3}$$
$$M_2 = C_{min}(Q_0 - Q_1)(e^{k_bt_B} - e^{k_bt_A}). \tag{2.4}$$

Substituting 2.2, 2.3, and 2.4 into 2.1 and simplifying, the booster efficiency for the first-order demand case becomes

$$\epsilon_T = e^{k_bt_A}\frac{Q_1}{Q_0}e^{-k_bt_J}(e^{-k_bt_A} - e^{-k_bt_B}) + (e^{-k_bt_A} - e^{-k_bt_B})(1 - e^{-k_bt_J}), \tag{2.5}$$

where ϵ_T is the "total" booster efficiency, so called because it includes the contributions of both the physical distribution effect and the chemical kinetics effect. The physical distribution effect will be isolated from this expression below.

In contrast to the first-order kinetic assumption, the constant demand assumption implies that the chlorine demand per unit volume between the source, junction, and points A and B is equal to that demanded for the conventional case (independent of changes in dose). As a result of this assumption, redistributing the dose between the clearwell and the booster station will have no effect on the total chlorine demand. Expressions M_1 and M_2, for the constant demand case, were developed by forcing the chlorine demand for the booster scenario equal to that for the conventional scenario,

$$M_1 = C_{min}Q_0[1 + e^{k_bt_B}(e^{k_bt_J} - e^{-k_bt_A})], \tag{2.6}$$
$$M_2 = C_{min}(Q_0 - Q_1)(e^{k_b(t_B-t_A)} - 1). \tag{2.7}$$

Substituting 2.2, 2.6, and 2.7 into 2.1 and simplifying, the booster efficiency for the constant demand case,

$$\epsilon_D = \frac{Q_1}{Q_0}e^{-k_bt_J}(e^{-k_bt_A} - e^{-k_bt_B}), \tag{2.8}$$

where ϵ_D is the booster efficiency attributable to the physical distribution effect, as the total chlorine demand per unit volume is fixed to that in the conventional case; thus, any reduction in chlorine use predicted by 2.8 must be due to better matching of dose rates to the physical distribution of the flow rates in the network. The chemical kinetic effect can be isolated from the total booster efficiency by subtracting 2.8 from 2.5,

$$\epsilon_K = (e^{k_b t_A} - 1)\frac{Q_1}{Q_0}e^{-k_b t_J}(e^{-k_b t_A} - e^{-k_b t_B}) + e^{k_b t_A}(e^{-k_b t_A} - e^{-k_b t_B})(1 - e^{-k_b t_J}). \quad (2.9)$$

Equations 2.5, 2.8, and 2.9 are more easily analyzed in terms of the following dimensionless variable groups:

$$0 \le F = Q_1/Q_0 \le 1, \quad (2.10)$$
$$0 \le D = e^{-k_b(t_J+t_B)} \le 1, \quad (2.11)$$
$$0 \le Z = t_A/t_B \le 1, \quad (2.12)$$
$$0 \le J = \frac{t_J}{t_J + t_B} \le 1. \quad (2.13)$$

In physical terms, the *flow factor* F equals the fraction of total flow distributed to low residence time Zone 1; the *decay factor* D equals the ratio of chlorine concentration remaining at point B to that at the WTP, under the first-order assumption; the *zone factor* Z equals the ratio of low to high zone residence times; and the *junction factor* J equals the ratio of the junction residence time to the maximum residence time.

Noting that $t_B/(t_J+t_B)+J = 1$, 2.5 is expressed in terms of these dimensionless variable groups,

$$\epsilon_T(F, D, Z, J) = D^{-Z(1-J)}[FD^J(D^{Z(1-J)} - D^{1-J}) + (D^{Z(1-J)} - D^{(1-J)})(1 - D^{(1-J)})]. \quad (2.14)$$

Similarly, 2.8 and 2.9 become,

$$\epsilon_D(F, D, Z, J) = FD^J(D^{Z(1-J)} - D^{(1-J)}), \quad (2.15)$$
$$\epsilon_K(F, D, Z, J) = (D^{-Z(1-J)} - 1)FD^J(D^{Z(1-J)} - D^{(1-J)}) + (1 - D^{(1-J)(1-Z)})(1 - D^J) \quad (2.16)$$

The boosting efficiencies ϵ_T, ϵ_D, and ϵ_K are now uniquely determined by the four dimensionless quantities F, D, Z, J.

Physical Distribution Effect

The physical distribution effect is a strong function of the distribution of flows between Zones 1 and 2. Distribution efficiency, ϵ_D, is shown in figure 2.6 (left) as a function of the zone factor Z (for $D = J = 0.1$), and increases significantly with the flow factor F. Indeed efficiency increases for all Z values as F increases from 0.1 to 0.9. When chlorine demand in Zone 1 is low (a short residence time, or small Z values) and the distribution of flow to Zone 1 is high (F is large) this effect is pronounced, since the disinfectant mass per unit volume required at the WTP for residual maintenance in Zone 1 is smaller relative to the

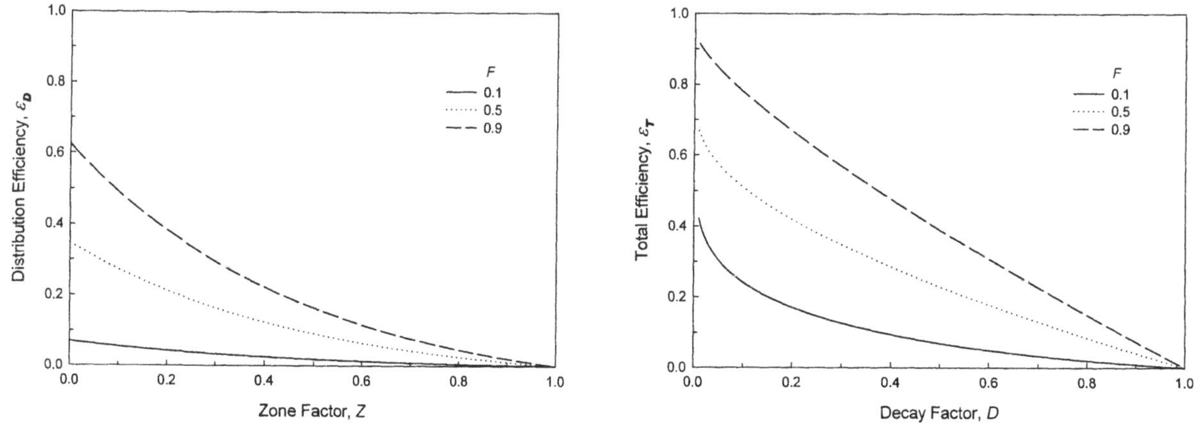

Figure 2.6: Distribution efficiency, ε_D, as a function of zone factor, Z (left: $D = 0.1$, $J = 0.1$, F as indicated). Total efficiency, ε_T, as a function of decay factor, D (right: $Z = 0.1$, $J = 0.1$, F as indicated.

conventional case, and a large proportion of the flow is being diverted there. Thus, high branch flow relative to total flow (large F values) represent favorable flow conditions for reducing chlorine use via booster disinfection.

As eluded to above, improvements in ϵ_D are also dependent on the relative chlorine demand between the WTP and point A, and that between the booster station and point B. As Z approaches 1, ϵ_D decreases to zero for any F, because as Z increases the chlorine demand per unit volume in Zone 1 approaches that of Zone 2. Thus the chlorine dose required to maintain residuals in Zone 1 also approaches that for Zone 2, and any advantage in boosting is eliminated. The effect of J on distribution efficiency is similar. As J approaches 1, $t_J + t_A$ approaches $t_J + t_B$ (recall that $t_A \leq t_B$), and so for reasons analogous to those for Z approaching 1, the advantage of boosting disappears.

Chemical Kinetic Effect

The kinetic efficiency ϵ_K is shown in Figure 2.7 (left) as a function of the junction factor J ($D = Z = 0.1$). The kinetic efficiency exhibits a maximum in its dependence on J and, for this small zone factor, is relatively insensitive to the flow factor F. The maximum in $\epsilon_K(J)$ can be explained intuitively by considering a limiting case in which F and Z both approach zero (as it makes no sense to require residual maintenance in a zone with zero demand). As F and Z approach zero, Zone 1 becomes of little consequence to the analysis, and the problem becomes one of determining the best location of a booster station for delivering a specified dose rate between the WTP and point B. From a chemical kinetic perspective, the optimal solution is to split the dose evenly between the WTP and the booster station, and locate the booster station midway between the WTP and point B (ie., $J = 0.5$). This solution minimizes the average chlorine concentration and, under the first-order assumption,

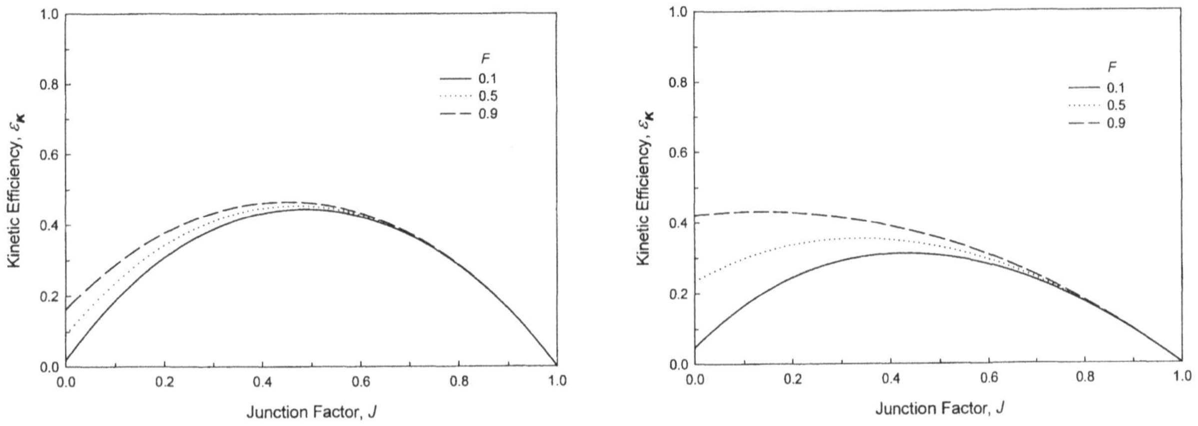

Figure 2.7: Kinetic efficiency, ε_K, as a function of junction factor, J (left: $D = 0.1$, $Z = 0.1$, F as indicated; right: $D = 0.1$, $Z = 0.5$, F as indicated).

minimizes the chlorine demand in the distribution system. Moving the booster location away from the midpoint will increase the demand in the longer travel time segment more than it is reduced in the shorter segment - hence the kinetic efficiency is maximized at $J = 0.5$, for this limiting case.

The general behavior of booster kinetic efficiency can be explained using arguments that draw from the $F = Z = 0$ limiting case. Since the WTP dose is responsible for maintaining residual in Zone 1 (time $t_J + t_A$), as Z increases from zero the optimal booster location will be at a junction closer to the WTP, when compared to the above limiting case (i.e., $J < 0.5$). Moving the location toward the WTP will reduce chlorine demand from the WTP to point A more than it increases the demand from the booster station to point B. The actual optimal location will, however, depend on the values of Z and F. When Z is non-zero but small, the travel time in Zone 1 remains a small fraction of that in Zone 2. Thus, the impact of Zone 1 on the optimal booster dose location is also small, as shown in figure 2.7, where the optimal kinetic efficiencies occur near $J = 0.5$ for any F.

As F increases, the chlorine demand per unit volume from the WTP to point A is multiplied by a larger flow rate, and hence reductions in this chlorine demand per unit volume are emphasized relative to that in the Zone 2 branch. As a result, when F increases the optimal kinetic efficiency is obtained by boosting at junctions closer to the WTP. When Z is small this effect due to F is limited, as shown in Figure 2.7. Yet, as the travel time in Zone 1 becomes a significant fraction of that in Zone 2 (increasing Z) this effect of a change in F can be substantial. Figure 2.7 (right) illustrates the kinetic efficiency for a case identical to that in Figure 2.7 (left), except the zone factor Z has been increased to 0.5. For these relatively longer travel times in Zone 1, the total chlorine demand is reduced by locating the booster station at a junction substantially nearer the WTP when compared to Figure 2.7 (left). In the beneficial case where the flow to Zone 1 predominates (e.g., $F =$

0.9), the optimal travel time to the booster station can be less than 20% of the maximum distribution system travel time.

Total Boosting Efficiency

A synergy of effects is present when conditions favorable to both the physical distribution and chemical kinetic effects are combined. Such a situation occurs for small values of D, as shown in figure 2.6 (right), which shows total efficiency as a function of D (with $Z = J = 0.1$). When flow conditions are favorable (high branch flow — large F) $\epsilon_T(D)$ can be in excess of 80% (for F=0.9 and $D < 0.15$), since both ϵ_D and ϵ_K have maximums which combine under these conditions. Starting with the extreme case when D=1, neither ϵ_D nor ϵ_K allow any advantage since there is no chlorine decay. As D approaches 0, however, a maximum in distribution efficiency, ϵ_D, occurs when D=0.1; under these conditions, the rate of disinfectant decay is such that the difference in the mass of chlorine demanded per unit volume between Zones 1 and 2 is significant relative to the conventional case. As D decreases further, however, distribution efficiency plummets as overall chlorine demand becomes so great as to make the difference in demand per unit volume between the zones negligible for any set of flow conditions. In contrast to distribution efficiency, kinetic efficiency ϵ_K increases exponentially over the entire range as D decreases and is maximized when D=0. As decay conditions become increasingly severe, the advantages of reducing the average concentration, and therefore, the average rate of decay become significant when considering kinetic efficiency. Practically speaking, when flow conditions are favorable, the physical distribution effect and the chemical kinetic effect combine to make booster chlorination the most efficient method for maintaining a residual in regions of the distribution system where it is difficult under conventional disinfection practices.

Interpreting First-Order Efficiency Estimates

Bulk chlorine decay is more complex than either of the limiting cases used to bracket it for the purposes of this analysis. Indeed, first-order bulk chlorine decay is the most severe decay expected under rechlorination conditions. For that reason, it is likely that first-order estimates of total boosting efficiency are optimistic. Nevertheless, for any case where $D > 0$, there exists both a chemical kinetic and a physical distribution effect. The boosting efficiency we can expect with a more realistic chlorine decay model will increase with F for any set of decay conditions represented by $D > 0$, $J > 0$, and $Z < 1$ and is bounded between the ϵ_D and ϵ_T estimates. Thus, relatively independent of the chlorine decay kinetics, chlorine mass can be saved by locating and scheduling the doses in harmony with the joint residence time and flow distribution in the network.

Chapter 3

Current State of Practice

This chapter reviews results of a survey on booster disinfection usage by United States water utilities, and reviews the known published case studies on booster disinfection. The booster disinfection survey was conducted in two phases. Phase I was a brief postcard survey to assess the frequency of use of booster disinfection. Phase II was a follow-up survey sent only to those utilities responding in the postcard survey that they currently used booster disinfection in their distribution system. Both survey instruments are included in Appendix B.

3.1 Survey of Booster Disinfection Use

Booster disinfection is not a new technology. A survey of utility disinfection practices performed in 1992 indicates that approximately 30% of the participating utilities applied disinfectants in the distribution system(AWWA, 1992b). The Phase I postcard survey of 4000 U.S. treatment plants indicates that 15% of the approximately 800 responding utilities currently practice booster disinfection. The most common reasons for practicing booster disinfection were: 1) distribution system residual maintenance, 2) biological regrowth prevention, and 3) disinfection after open reservoirs.

Phase II surveys were sent to 122 utilities who indicated they practice booster disinfection; 38 utilities responded to the phase II survey, while 8 additional surveys were returned without being filled out because the utilities did not practice booster disinfection. These 38 completed surveys were analyzed statistically.

The utilities practicing booster disinfection ranged from .14 to 820 MGD, with an average of 55 MGD. Approximately 2/3 applied pre- and post-oxidant, while 1/3 applied post-oxidant only. The most common pre-oxidant was chlorine gas (58%), followed by potassium permanganate (17%) and hypochlorite solution (17%), with similar percentages for post-oxidant. Approximately 12% of utilities used chloramine distribution system residual.

The most common distribution system problems reported by these utilities were low disinfectant residual (70%), stagnant water (40%), discoloration (30%), unpleasant tastes and odors (26%), and high DBP formation (20%). These utilities operated booster stations that ranged in age from new to 78 years old, with an average age of 16 years. The purpose of the booster stations was overwhelmingly to solve low residual problems (85%), followed by "other" (27%) and biological regrowth (15%). On a scale of satisfaction with booster disinfection from 1 (not satisfied) to 5 (very satisfied), the average rating was 4.1, with a low of 2 and a high of 5. Thus it appears that utilities using booster disinfection are fairly satisfied with their performance, for those problems they were intended to solve.

The 57 booster stations that were reported on (some utilities reported on more than one station) were installed at locations where the flow is somewhat variable (average of 3 on a scale of 5), and on average are set to deliver a chlorine concentration of 2.3 mg/L. The siting of stations was due mainly to proximity to storage tanks (54%) or pump stations (33%), followed by proximity to areas with low residual (17%). One presumes, however, that siting close to a storage tank or reservoir is essentially to mitigate a low residual problem. Somewhat surprising is that 25% of utilities used distribution system hydraulic or hydraulic/water quality simulation models to assist in the siting decision. Apparently safety concerns about chlorine gas are thought to be manageable by the utilities, since 60% of the booster stations use chlorine gas, while 30% use hypochlorite solution. This may reflect the relative convenience of using chlorine gas, compared to maintaining hypochlorite solution on a regular basis. Interestingly, 3 booster stations were feeding both chlorine and ammonia at the same location.

To monitor residuals 20% monitored just upstream of the booster injection, 28% just downstream, and 56% in the distribution system zone that is affected by the booster station. This result suggests that most utilities are regulating their booster station dosages to control residuals far away from the station, out in the distribution system (presumably these utilities are also monitoring close to the station for local process control). In terms of monitoring frequency, 14% used continuous analyzers, 24% took daily grab samples, 25% took weekly grab samples, and 16% only sampled monthly.

The installed booster stations seem mostly to deliver a constant disinfectant mass rate (55%), but 38% are using some type of feedback or feedforward control to attempt to maintain a constant "set-point" concentration (either through continuous analyzers or using flow-pacing). Only rarely (7%) are utilities attempting to maintain a time-varying set-point concentration, but this technology has been implemented at several stations. Most (57%) booster stations are controlled manually on-site, though one-third are regulated via automatic controls, and 10% by remote (SCADA) manual control. Of those utilities implementing automatic control, only 2.8% use redundant sensors, which may mean that utilities are choosing to check/calibrate sensors manually rather than through redundancy. Almost half the automatic control booster stations use remote alarms, while 14% use automatic high/low dose cutoffs, and 11% use other "fail-safe" features.

3.2 Booster Disinfection Literature Review

Relative to the frequency of it's practice, little research has been performed measuring the ability of booster disinfection to control biological regrowth, or its affect on DBP formation. Several case studies have, however, explored the overall benefits of booster disinfection. Edge and Finch(Edge and Finch, 1987) performed a case study to evaluate the ability of booster disinfection to prevent the growth of Aeronomas hydrophilia, an organism that may be related to human health effects. The system consisted of a small village which received treated, and chlorinated, surface water from a water tower, and an inner city which received a blend of treated, and chlorinated, surface and ground water from a storage reservoir. During the first year of the study, a booster chlorination station was placed at the outlet of the water tower. This single booster station provided adequate residuals and Aeronomas hydrophilia inactivation in the vicinity of the water tower, however, the residuals decayed quickly and Aeronomas hydrophilia were still detected at more remote sites of the village. In an attempt to provide a longer lasting residual, the booster station was switched to chloramination, however, this too only provided adequate residual and bacterial removal in the vicinity of the water tower. In the inner city, residual maintenance and inactivation of Aeronomas hydrophilia was accomplished by increasing the chlorine dose leaving the treatment plant and chlorine dosing at the storage reservoirs. Over the winter of the first year, booster stations were installed at all storage locations, and the pipe configuration was rearranged to provide the water tower (and the village) with a 50%/50% blended surface/ground water supply. These changes were able to provide a 0.2 mg/L residual throughout the distribution system and eradicate Aeronomas hydrophilia. Additionally, there was very little difference in THM formation for the unboosted (77 g/L) and boosted (82 g/L) water. The benefits observed in the village, however, could not be entirely attributed to the addition of booster chlorination as the blending of the surface and ground water may have improved the overall water quality in general.

Gibbs et al.(Gibbs et al., 1990) investigated the differences in microbiological activity and THM formation between two different zones of a distribution system, one without (A) and one with (B) booster chlorination. The largest residence times are greater than 3 days. The water supplied to these zones come from separate storage reservoirs (both reservoirs receive treated, and chloraminated, surface water from the same treatment plant). During the first year neither zone was booster chlorinated to determine the extent of any correlation in water quality parameters. Heterotrophic plate count (HPC) was observed to be correlated between the two reservoirs. During the second year, the effluent of the reservoir feeding zone B was booster chlorinated. It was observed in zone B, via ten sampling points, that the disinfectant residual and HPC were judged acceptable closer to the storage reservoir, but degraded with distance from the reservoir. The authors suggest the boosted chlorine increases the assimilable organic carbon, thus increasing the microorganism food supply. This is a chloraminated system, however, so it is possible the chlorine applied moved the disinfectant residual into the breakpoint chlorination range, thus decreasing the effective disinfectant concentration. Unfortunately the combined residual was not reported. As with the study by Edge and Finch(Edge and Finch, 1987), Gibbs et al. showed little increase in THM formation from the conventionally treated (66 g/L) to the boosted (72 g/L) water.

A case study by Devkota et al. involved a wastewater treatment plant that used booster chlorination in a chloramine system. Their study goal was to reduce chemical costs by decreasing the required chlorine dose. They compared single and staged dose scenarios for a 60 minute detention time. Unchlorinated secondary effluent samples were initially dosed with enough stock ammonia solution to achieve a concentration of 0.5, 1.0, or 1.5 mg NH_3-N/L with an associated chlorine concentration of 5, 10, or 15 mg Cl_2/L. Under the single dose scenario, all chlorine was applied initially. During the staged dose scenario 50% of the single dose was applied initially, followed by 33% after 30 minutes of decay, and 17% after 20 minutes. While the authors indicate they wish to avoid Cl:N ratio greater than 7.6:1, the ratio for the single dose was 10:1. For the two lower concentration doses, staged dosing provided a 0.3 - 0.4 mg Cl_2/L higher residual than the respective single dose.

Kiene et al.(Kiene et al., 1998) described a booster chlorination case study in the French town of Cholet which serves approximately 60,000 inhabitants. The goals of the booster stations were to increase the percentage of the distribution system achieving a detectable residual, and reduce the amount of chlorine applied. Two booster stations were selected for their ability to increase the disinfectant residual coverage (these locations were selected using a distribution system hydraulic/water quality model). The booster stations increased the residual coverage in the system from 35% (pre-booster stations) to 49%. The stations also reduced THM formation from 80 $\mu g/L$ (with a 0.60 mg Cl2/L residual leaving the treatment plant) to 65 $\mu g/L$ (with a 0.40 mg Cl2/L residual at the treatment plant and an additional 0.25 mg Cl2/L at the booster stations). The total disinfectant mass applied decreased using booster chlorination, although the magnitude of reduction could not be calculated from the data supplied. The THM concentration did not increase after booster chlorination, while HPC and Aeronomas hydrophilia levels were an order of magnitude less in areas influenced by booster chlorination, compared to conventional chlorination.

Ireland and Knudson(Ireland and Knudson, 1998) reported that the Portland Water Bureau has practiced booster chlorination of this chloraminated system since 1983. The system primarily uses two booster chlorination stations located at the effluents of reservoirs, and have recently added a temporary booster chlorination station to an area of the distribution system that can experience low residuals and nitrification. The primary objectives of these chlorination stations are to combine with ammonia, the energy source of the nitrifying bacteria, and to increase the chloramine residuals (of which they focus on maintaining primarily monochloramine). The booster stations have been shown to initially reduce the ammonia concentration, however, the concentration begins to increase again as chloramine decay continues. Overall, Ireland and Knudson believe that booster chlorination has been “a major factor” in providing an adequate residual and maintaining water quality.

Sohn et al.(Sohn et al., 1997) evaluated DBP formation in the city of Paris distribution system which practices booster chlorination. The study used power function correlation models to compare in-plant and distribution system levels of TTHM formation and predicted distribution system TTHM yields with a maximum error of 40%. Sohn cited a number of confounding factors that made the comparison of these TTHM levels difficult, including: residence time and chlorine concentration variability; changing reactivity of natural organic matter before and after chlorination; absence of pH and temperature measurements, and the unknown effects of biofilm, storage tanks, and wall reactions that consume free chlorine.

Chapter 4

Chlorine Decay and THM Formation

This chapter focuses on bulk disinfectant decay and disinfection by-product (DBP) formation under booster disinfection conditions. The specific objectives are:

1. Evaluate bulk decay of free chlorine and DBP formation as a function of initial concentrations, boost dose, boost time, and pH, for various treated waters.

2. Develop a mathematical model to describe bulk chlorine decay and DBP formation under booster disinfection conditions, and compare to the typical first order kinetic model.

In addition to these primary objectives, preliminary insights into the effectiveness of booster disinfection for chloraminated distribution systems are discussed.

4.1 Background

Chlorine and chloramine are the two most common disinfectants in the United States, with chlorine being the predominant disinfectant. The primary purpose for chlorine addition is the inactivation of microorganisms. Secondary purposes are to bleach color causing components, to oxidize metals or compounds that impart bad tastes and/or odors, to pre-oxidize organic matter to improve coagulation, to provide a secondary barrier to pathogens entering the distribution system via cross-contamination, and to prevent regrowth of microorganisms within the distribution system. This latter objective is also the primary use of chloramines. Unfortunately these disinfectants also decay within the distribution system by reacting with material both in the bulk matrix and/or at the pipe wall surface to form various DBPs. Some DBPs are suspected carcinogens(Bull and Kopfler, 1991), and epidemiological and animal studies have suggested a link between certain DBPs and pregnancy loss(Waller et al., 1998; Bielmeier et al., 2001). The number and type of DBPs are quite extensive(Clark et al., 1994;

Singer, 1994) and depend on disinfectant type. Chlorination produces more halogenated DBPs than chloramination. Krasner et al.(Krasner et al., 1989) found that the largest groups of halogenated DBPs are the trihalomethanes (THMs), while less is known about the nitrogenous DBPs of chloramine. In the following sections we review basic chlorine and chloramine chemistry and what fundamentals are known of the reactions between chlorine and natural organic matter (NOM). We also review the various factors affecting chlorine and chloramine decay, and THM formation duc to chlorination for inactivation or residual maintenance. Finally, models which have been proposed to predict chlorine decay and THM formation are discussed, as background for the models developed in section 4.7.

4.1.1 Basic Chlorine and Chloramine Chemistry

Chlorine is commonly available in a gaseous form at normal temperature and pressure (Cl_2) and in a solid or liquid form (NaOCl or $Ca(OCl)_2$). Upon addition to water Cl_2 undergoes a hydrolysis reaction,

$$Cl_2 + H_2O \rightarrow HOCl + H^+ + Cl^- \tag{4.1}$$

and, for example, NaOCl dissociates

$$NaOCl + H^+ \rightarrow HOCl + Na^+, \tag{4.2}$$

converting the non-disinfecting chlorine compounds to hypochlorous acid, HOCl(White, 1992). HOCl is a weak acid, thus it will partially dissociate to give

$$HOCl \leftrightarrow H^+ + OCl^- \tag{4.3}$$

with a pKa of 7.5 at 25°C(Snoeyink and Jenkins, 1980). At pH < 7.5, HOCl is the predominant species although for pH < 2 Cl_2 gas is liberated, as shown in Figure 4.1. At pH > 7.5, however, OCl^- is the predominant species. Additionally, HOCl is a much better disinfectant than OCl^-, and thus the disinfectant ability of chlorinated water is much lower at pH 9(White, 1992). Since drinking water pH values are typically between 6 and 9, the impact of pH on disinfectant ability is significant for achieving microorganism inactivation.

Reaction mechanisms for the formation of THMs have not been well characterized as the heterogeneous structure of NOM is poorly understood. Most kinetic work related to reaction mechanisms of THM formation has been done using model organic compounds, those compounds believed to have similar functional groups to NOM. Morris and Baum(Morris and Baum, 1978) presented a set of reaction mechanisms that is generally accepted to describe THM formation. This reaction involves enolization (the base promoted halogenation of a ketone(Morrison and Boyd, 1987)) of a methyl ketone which undergoes halogenation to produce a monohalogenated compound (see Figure 4.2). The product of thes reactions may proceed through this process two more times, producing a trihalogenated compound which undergoes hydrolysis to form a THM (see Figure 4.3). This process is base catalyzed and

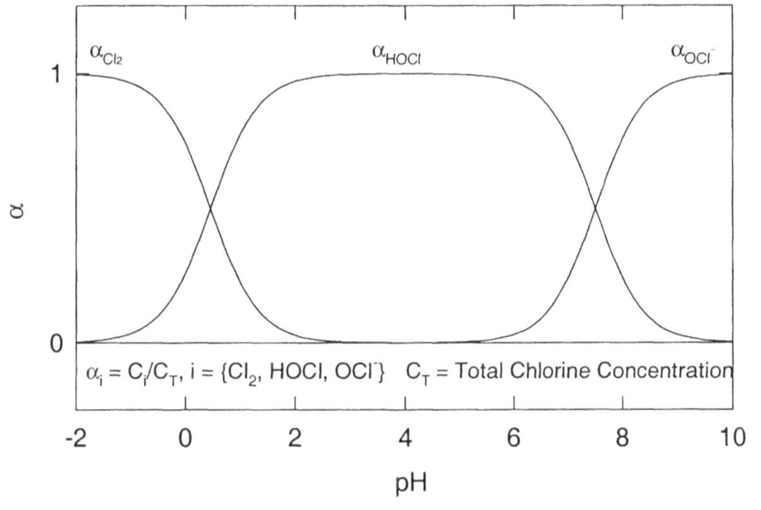

Figure 4.1: Chlorine speciation as a function of pH.

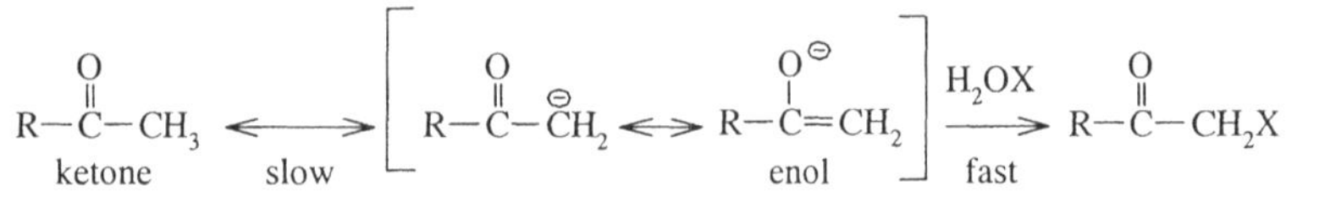

Figure 4.2: Reaction mechanism for monohalogenated DBP.

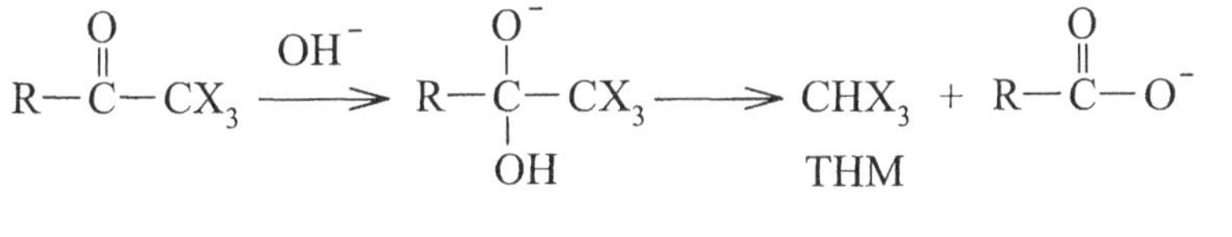

Figure 4.3: Reaction mechanism for trihalogenated DBP.

thus enhanced under typical finished drinking water conditions, as the pH is typically in the basic range.

Hypochlorous acid can react with ammonia, either naturally present in the water or added intentionally, to produce chloramines. The reaction occurs by,

$$HOCl + NH_3 \leftrightarrow NH_2Cl + H_2O, \tag{4.4}$$

which produces monochloramine (NH_2Cl). This reaction occurs fastest at a pH near 8.3 and slows down with either increasing or decreasing pH(White, 1992). Additional reactions can occur with HOCl to produce dichloramine and trichloramine. One set of reactions is:

$$NH_2Cl + HOCl \leftrightarrow NHCl_2 + H_2O, \tag{4.5}$$

and trichloramine,

$$NHCl_2 + HOCl \leftrightarrow NCl_3 + H_2O. \tag{4.6}$$

These reactions, however, should not normally produce large concentrations of di- or trichloramine within a treatment plant as 4.5 and 4.6 require chlorine to nitrogen ratios (Cl:N) $>$ 5:1, by weight, and pH < 6 to produce significant concentrations of $NHCl_2$ and NCl_3(White, 1992; Bryant et al., 1992; Norman et al., 1980), and neither condition are typical in drinking water. Diehl et al.(Diehl et al., 1995) showed that at pH 6, after 48 hours of reaction time, dichloramine accounted for 30 (Cl:N = 3:1) to 70 (Cl:N = 7:1) percent of the total combined chlorine residual, which was not present in significant amounts at pH 8 or 10.

Chloramines decompose even in the absence of other reactive material. Stoichiometrically, (4.4) requires a Cl:N of 5:1 (1:1 by molar) for complete reaction; increasing the Cl:N ratio beyond 5:1 begins to produce dichloramine (4.5). At typical drinking water pH, however, dichloramine degrades by

$$2NHCl_2 + H_2O \rightarrow N_2 + HOCl + 3H^+ + 3Cl^-, \tag{4.7}$$

which reduces the chlorine strength due to the reduction of chlorine, Cl^{+1}, to chloride, Cl^-, which is not a disinfectant(Bryant et al., 1992). This occurs, theoretically, until a Cl:N of 7.6:1 is achieved, at which point all of the ammonia-nitrogen will be oxidized to molecular nitrogen. After the complete oxidation of the ammonia-nitrogen, the addition of more chlorine will produce a free chlorine residual. This phenomena is also known as the breakpoint curve and is shown in Figure 4.4. At Cl:N $<$ 5:1, the redox reactions occur slower as the chlorine is no longer in excess, thus little, if any, $NHCl_2$ is formed. At Cl:N $<$ 5:1 monochloramine decomposes by

$$3NH_2Cl \rightarrow N_2 + NH_3 + 3H^+ + 3Cl^- \tag{4.8}$$

which yields a net increase in NH_3 concentration that increases in the presence of organic matter(Valentine et al., 1997). The production of NH_3 in (4.8) may lead to nitrification - the biological oxidation of ammonia (NH_3) to nitrite (NO_2^-) by *Nitrosomonas* (most common

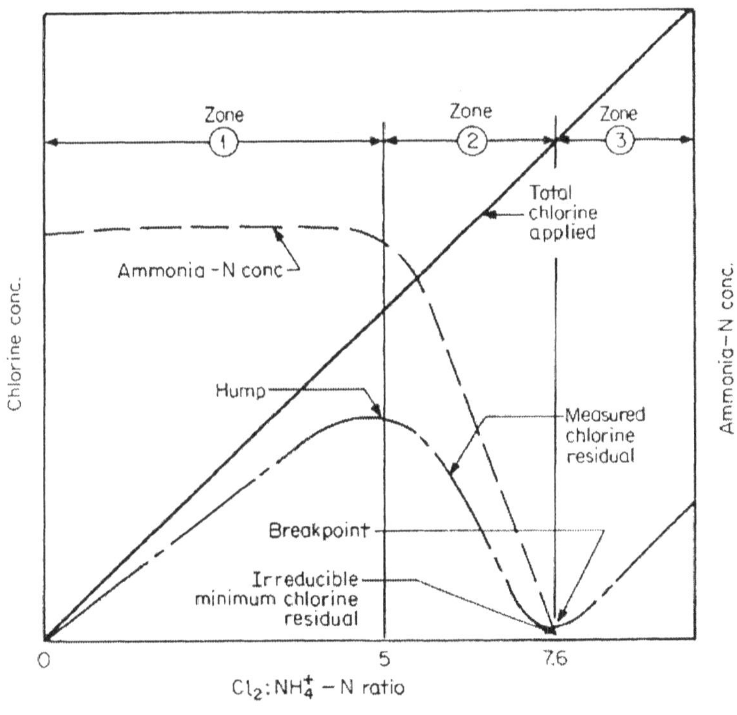

Figure 4.4: Theoretical breakpoint curve of chlorine with ammonia(White, 1992)

organism) followed by the biological oxidation of nitrite to nitrate (NO_3^-) by *Nitrobacter* (most common organism)(Rittman and Snoeying, 1984). While complete nitrification has been suggested to promote biologically stable water(Rittman and Snoeying, 1984), partial nitrification, the first oxidation step ($NH_3 \rightarrow NO_2^-$), may cause loss of disinfectant residual and increased heterotrophic plate counts (HPC)(Wolfe et al., 1988). There are undesirable health effects associated with nitrates and nitrites, which are regulated by the U.S.E.P.A. at a maximum contaminant level (MCL) of 10 ppm and 1 ppm, respectively, in drinking water.

Partial nitrification occurs due to a competing chemical oxidation of NO_2^- to NO_3^- by HOCl. NO_2^- has a large, and rapid, chlorine demand (1 mg N/L reduces 5 mg Cl_2/L(White, 1992)), thus the small amounts of HOCl present under chloramination quickly oxidizes NO_2^- to NO_3^-. The biological removal of NH_3 and the chemical reduction of HOCl cause the equilibrium reactions in (4.4-4.6) to shift to the left forming more HOCl and NH_3, which allows these processes to proliferate. In turn, this decreases the chloramine concentration and may explain why chloramine residuals are difficult to maintain during nitrification episodes (also suggested by Mann et al.(Mann et al., 1998)).

When HOCl is present with bromide it will react very quickly to form hypobromous acid, a weak acid with a pKa = 8.7(White, 1992),

$$HOCl + Br^- \rightarrow HOBr + Cl^- \tag{4.9}$$

thus reducing the amount of free chlorine and the associated disinfectant ability of chlorine.

HOBr is a less effective oxidant than HOCl, yet a more reactive halogenating agent. Therefore HOBr decreases inactivation while increasing the rate of THM formation(Larson and Weber, 1994). In the presence of chloramine, a similar reaction to (4.4) occurs in which monochloramine forms monobromamine

$$NH_2Cl + Br^- \rightarrow NH_2Br + Cl^- \tag{4.10}$$

which also undergoes decomposition similar to, yet faster than, the decomposition of monochloramine. Additionally, bromochloramines may form, along with the bromamines, and may increase the overall decay of chloramines(Valentine et al., 1997).

HOCl and OCl^- are both stronger disinfectants than NH_2Cl(White, 1992; Bryant et al., 1992), which gives chlorine an advantage for use as a primary disinfectant, since chlorine requires either a shorter contact time or lower dose to produce the same level of inactivation as chloramine(Norman et al., 1980). For secondary disinfection, required to maintain residuals and prevent biological regrowth, chloramines have been shown to provide more stable residuals(Bryant et al., 1992; Norman et al., 1980), and a more biologically stable water(Bryant et al., 1992) via plate counts(Neden et al., 1992; Norman et al., 1980) and positive coliform tests(Norman et al., 1980). Although dichloramine has been reported to be a stronger disinfectant than monochloramine(White, 1992), it produces taste and odor problems at very low concentrations. Trichloramine has also been used as a strong oxidant, however its use as a disinfectant has not been studied and in any case it has worse taste and odor problems at low concentrations than does dichloramine(White, 1992).

Since chlorine is a very strong oxidant it can be used not only for disinfection, but to aid in the oxidation of iron and manganese, destruction of taste and odor causing compounds, and bleaching of color causing compounds. Additionally, due to the relative non-specific reactivity of chlorine, halogenated disinfectant by-products are also formed by reaction with the organic material in the water. Chloramines, on the other hand, are much less reactive, which is believed to result from the more electronegative center of NH_2Cl (N^-3), as compared to HOCl (O^-2), binding more strongly with chlorine (Cl^+1)(Bryant et al., 1992). This decreased reactivity results in fewer halogenated DBPs(Stevens et al., 1989), however, other non-halogenated by-products may be formed, of which less is known.

4.1.2 Factors Affecting Disinfectant Decay and THM Formation

Several factors can contribute to disinfectant decay and THM formation including Cl:N ratio, disinfectant dose, inorganics, pH, temperature, and organic matter. Each factor will be discussed individually in regard to its effect on disinfectant decay and THM formation, all else being equal.

Chlorine to Nitrogen Ratio

Studies by Valentine et al.(Valentine et al., 1997), Jafvert and Valentine(Jafvert and Valentine, 1992), and Lieu et al.(Lieu et al., 1993) indicate that increasing the Cl:N increases NH_2Cl decay. This increase in decay is due to excess chlorine forming dichloramine, which then decays rapidly (Eq. 4.7). Symons et al.(Symons et al., 1996b) suggest that THM formation remains relatively low at Cl:N less than 5:1, increasing significantly at Cl:N > 5:1, which is assumed related to the more reactive dichloramine. Bryant et al.(Bryant et al., 1992) stated that THM formation increases as Cl:N approached the breakpoint Cl:N of 5:1. Diehl et al.(Diehl et al., 1995) evaluated THM formation for Cl:N of 3:1, 5:1, and 7:1 and showed no difference in THM formation at pH 10, increasing THM formation with increasing Cl:N at pH 8, and increasing THM formation at pH 6 with Cl:N of 5:1 > 7:1 > 3:1, which is suggested as "evidence of some complexity in the haloamine chemistry." Bryant et al.(Bryant et al., 1992) and Wolfe et al.(Wolfe et al., 1988) suggest that increasing the Cl:N could reduce the concentration of ammonia available for nitrifying bacteria thus prohibiting the onset of nitrification. Lieu et al.(Lieu et al., 1993) showed that increasing the Cl:N did, to some degree, inhibit the growth of nitrifying bacteria.

Dose

Fleischacker and Randtke(Fleischacker and Randtke, 1983) showed that increasing the chlorine dose increased decay rates. They evaluated several initial chlorine doses between 2 mg/L and 100 mg/L at pH 7 and a total organic carbon concentration (TOC) of 3.0 mg/L with the resulting demands ranging from 2 mg/L to 30 mg/L after 100 hours of reaction time, indicating that chlorine decay rate depends on disinfectant dose.

Fleischacker and Randtke(Fleischacker and Randtke, 1983) also observed an increase in THM formation from 105 g/L to 193 g/L with the increase in chlorine dose. Symons et al.(Symons et al., 1996a) and Stevens et al.(Stevens et al., 1989) both showed increases in THM formation with increases in chlorine dose. Symons et al.(Symons et al., 1996a) evaluated the THM formation at a pH of 8.2 and 25°C after 7 days of reaction time. Stevens et al.(Stevens et al., 1989) showed increases in THM formation with increases in chlorine dose from 8.1 mg/L to 20 mg/L at pH values from 5 to 11.5.

Lieu et al.(Lieu et al., 1993) showed little difference in chloramine demand with an increase in chloramine dose from 1.7 mg/L to 2.5 mg/L for Cl:N ratios of 3:1, 4:1, and 5:1. Fleischacker and Randtke(Fleischacker and Randtke, 1983) showed increased demand of 0.5 mg/L to 8.5 mg/L after 100 hours of reaction time for chloramine doses of 2 mg/L and 20 mg/L respectively, although the Cl:N of the doses were not provided. From the Fleischacker and Randtke study, the demand resulting from the 20 mg/L chloramine dose is larger than the demand from the 20 mg/L chlorine dose (demand = 5 mg/L), which was attributed to the increased reactivity of dichloramine. Fleischacker and Randtke also showed an increase in THM formation from < 1 g/L to 28 g/L with increased chloramine dose. There was not, however, a significant amount of THM formation until the initial chloramine dose exceeded 10 mg/L, a value not expected in practice.

Inorganics – Bromide

For both surface and ground water extracted humic acids with a TOC of 4.0 mg/L, an initial chlorine concentration of 8 mg/L and pH 7.0, Cowman and Singer(Cowman and Singer, 1996) showed that by increasing the bromide concentration from 0 to 2.0 mg/L, the chlorine consumed after 24 hours also increased by approximately 2 mg/L. Krasner et al.(Krasner et al., 1996), Symons et al.(Symons et al., 1994; Symons et al., 1987) showed that increasing the bromide concentration increased the THM formation. Krasner et al.(Krasner et al., 1996) showed a 210% to 260% increase in THM formation after 3 hours of decay when increasing the bromide concentration from 0.005 mg/L to 0.8 mg/L for 0.45μm filtered San Francisco Bay/Sacramento-San Joaquin River Delta water with DOC concentrations ranging from 1.10 mg/L to 4.15 mg/L, pH of 8.2, and 25C. Symons et al.(Symons et al., 1994) showed an 89% increase in 4 day THM formation of treated Lake Houston water by increasing the bromide concentration from 0.07 mg/L to 0.53 mg/L. Symons et al.(Symons et al., 1987) showed anywhere from 10% - 200% increases in THM formation with bromide concentrations ranging from 0 - 1.3 mg/L for 5 mg/L of commercially available soil humic acid, an initial chlorine concentration of 10 mg/L, reaction times of 4 hours to 6 days, pH range of 6 - 11, and 25C.

Valentine et al.(Valentine et al., 1997) showed that increasing the bromide concentration from 0.1 mg/L to 3.0 mg/L increased chloramine decay without bromide by 0.1 mg/L to 2.0 mg/L in the absence of any other reactive material at pH 7.5, 25C, a Cl:N of 3:1 and an initial monochloramine concentration of 3.5 mg/L. This additional decay between chloramine and bromide can be described by the reaction mechanisms presented by Valentine et al.(Valentine et al., 1997). Diehl et al.(Diehl et al., 1995) showed the addition of 0.5 mg/L bromide at pH 6 more than doubled the THM formation for Cl:N of 3:1, 5:1, and 7:1 after 48 hours of reaction time, and is suggested to be a function of the increased reactivity of dichloramine. This effect enhancing THM formation decreases with increasing pH until pH 10, at which point bromide addition has no effect on THM formation. This observation is consistent with the lack of dichloramine formation at higher pH.

Inorganics – Iron, Manganese, and Hydrogen Sulfide

Ferrous iron (Fe^{+2}) and hydrogen sulfide (H_2S) are both oxidized almost instantaneously by HOCl(White, 1992; Trussell and Umphres, 1978). The reaction with iron can be used to either remove iron or to create a coagulant, as oxidation of ferrous iron to ferric iron quickly forms an insoluble hydroxide(White, 1992). Manganese is also oxidized by chlorine to an insoluble oxide, and may require anywhere from 2 to 17 hours to react completely depending on the initial manganese concentration and pH(White, 1992; Shorney and Knocke, 1996).

Ferrous iron is also readily oxidized by chloramines (White, 1992) while the oxidation of manganese with chloramines is considerably slower than with chlorine (White, 1992; Shorney and Knocke, 1996). These comments about disinfectant/inorganic reactions assume the soluble, non-complexed inorganic species; if the inorganic species are complexed, the reactions may take significantly longer and may be unpredictable(White, 1992). Trihalomethanes are not formed by oxidation of iron or manganese with either chlorine or chloramines.

pH

Fleischacker and Randtke(Fleischacker and Randtke, 1983) and Koechling et al.(Koechling et al., 1997) both showed slight increases in chlorine decay rate with increasing pH. Fleischacker and Randtke showed an increase in chlorine demand of 0.4 mg/L when the pH was increased from 5 to 10 with an initial chlorine concentration of 3.0 mg/L and TOC of 3.2 mg/L. Koechling et al. showed an increased demand of about 1.0 mg/L when the pH was increased from 5.5 to 10 with an initial chlorine dose of 4.5 mg/L and DOC = 2.2 mg/L.

Fleischacker and Randtke(Fleischacker and Randtke, 1983), Stevens et al.(Stevens et al., 1989), Summers et al.(Summers et al., 1996), and Symons et al.(Symons et al., 1987) all showed increases in THM formation with increasing pH. This is consistent with the base promoted reaction presented by Morris and Baum(Morris and Baum, 1978). Fleischacker and Randtke showed a 290% increase in chloroform with increasing pH from 5 to 10. The work by Stevens et al. and Symons et al. covered chlorine doses over a range of 8.1 mg/L to 20 mg/L with increases in THM formation of 50% - 125% when increasing the pH from 5 to 10. Summers et al. showed a 50% - 60% increase in THM formation over a pH range of 6.75 - 8.75 while applying enough chlorine to provide a 1.0 mg/L residual after 24 hours.

Fleischacker and Randtke(Fleischacker and Randtke, 1983) showed no difference in chloramine decay over a pH range of 5 - 10 after 24 hours of decay. Jafvert and Valentine(Jafvert and Valentine, 1992), with an initial chloramine dose of approximately 45 mg/L, Cl:N $<$ 5 mg/L, and a pH range of 6 - 8, showed that decreasing the pH increased the decay of monochloramine due to the increased formation of dichloramine. For Cl:N $>$ 5:1, Jafvert and Valentine showed monochloramine decay increased by decreasing the pH from 8 to 7. Decreasing the pH to 6 actually decreased the monochloramine decay, although the initial concentration of monochloramine formed was also decreased due to the increase in dichloramine formation. Valentine et al.(Valentine et al., 1997) showed that for a chloramine concentration of 3.5 mg/L, decreasing the pH increased the decay of chloramine, regardless of the presence of organic material.

Fleischacker and Randtke(Fleischacker and Randtke, 1983) did not observe any chloroform formation under chloramination at any pH level. Both Diehl et al.(Diehl et al., 1995) and Stevens et al.(Stevens et al., 1989) observed decreasing THM formation with increasing pH, which is believed to be a result of the decrease in dichloramine formation at higher pH values.

Temperature

Koechling et al.(Koechling et al., 1997) showed that increasing the temperature from 11C to 34C increased the chlorine demand after 120 hours from approximately 2.5 mg/L to 4 mg/L with an initial chlorine concentration of 4.4 mg/L and a DOC of 2.1 mg/L. Fleischacker and Randtke(Fleischacker and Randtke, 1983) showed that increasing the temperature from 3C to 30C increased the chloroform concentration from 75 $\mu g/L$ to 230 $\mu g/L$ after 100 hours of contact time with an initial chlorine concentration of 20 mg/L and TOC of 3.0 mg/L. Summers et al.(Summers et al., 1996) showed that increasing the temperature from < 10C to > 30C increased THM formation 75 - 100% for three different source waters, while maintaining a 1 mg/L residual after 24 hours.

Valentine et al.(Valentine et al., 1997) showed approximately a 0.8 mg/L increase in chloramine demand after 150 hours when the temperature was increased from 4C to 35C with an initial chloramine concentration of 3.5 mg/L and no organic material present. Lieu et al.(Lieu et al., 1993) showed increases in chloramine demand after 28 to 42 days ranging from 0.4 to 1.2 mg/L with temperatures increasing from 10C to 25C and an initial chloramine concentrations of 1.7, 2.0, and 2.5 mg/L and Cl:N of 3:1, 4:1, and 5:1. Fleischacker and Randtke(Fleischacker and Randtke, 1983) showed no chloroform formation over 3C to 30C with an initial 20 mg/L chloramine dose allowed to react for 100 hours.

Total Organic Carbon

Laboratory studies using various raw and treated water sources have shown a positive correlation between chlorine demand and TOC(Summers et al., 1996), and between THM formation and TOC(Edzwald et al., 1985; Singer et al., 1981; Singer and Chang, 1989; Summers et al., 1996; Krasner et al., 1996; Symons et al., 1996a; McGuire et al., 1991). Since NOM is a heterogeneous mixture of various organic compounds, attempts have been made to relate gross NOM characteristics such as molecular size, hydrophobicity, and humic/non-humic fractions to THM formation in chlorinated waters(Sinha et al., 1997), and to chloramine demand(Valentine et al., 1997). Little, if any, work has been considered the effect of TOC or NOM characteristics on THM formation in a chloraminated system.

Pipe Material and Condition

The term "wall demand" will be used here to mean the reactions which occur between the disinfectant and various inorganic and organic material which might be located on the surface of the pipe wall. Wable et al.(Wable et al., 1991) examined the effects of chlorine decay in an actual distribution system at three different sites and compared the results to traditional bottle tests to determine the magnitude of wall demand. One site was supplied by a ground water source and exhibited a negligible difference in chlorine decay indicating little wall decay. The other two waters, fed by treated Seine river water, showed more decay in the pipe than the bottle tests indicating the wall material, or substances attached to the wall material, exerted an additional chlorine demand. Since the results compare the difference of chlorine decay in the distribution system (bulk and wall decay) to bottle tests (bulk decay only), the qualitative results should measure the influence of the wall decay component, assuming that reaction time in the distribution system could be measured accurately.

Brereton et al.(Brereton et al., 1997) examined THM formation in a batch reactor made of pipe wall material, and proposed a methodology for performing such experiments(Brereton and Mavinic, 1997). Their work indicated pipes, removed from a distribution system, contain adsorbed organic material and were shown to generate chloroform when the pipe reactors were filled with chlorinated organic free water. Using treated source water, chloroform production was less than the equivalent bottle tests for chlorine doses ≤ 4 mg/L, while for a chlorine dose ≥ 8 mg/L the THM formation was greater than the equivalent bottle test. The additional THM formation is suggested to occur from chlorine doses which are sufficiently large to overcome the demand from the pipe wall material, which allows the additional chlorine to react with the adsorbed organic material, thus forming more THMs.

4.1.3 Chlorine Decay and THM Formation Modeling

The multiobjective nature of disinfection, resulting from the tradeoff of disinfection and treatment benefits with its side effects, has provided greater motivation to describe chlorine decay and by-product formation through mathematical models. The following is a brief review of recent studies undertaken to describe chlorine decay and trihalomethane (THM) formation, a group of regulated disinfection by-products.

Describing both the rapid and slow chlorine decay with a consistent set of mechanisms has proved difficult. Lyn and Taylor(Lyn and Taylor, 1993) developed an empirical decay model for chlorine concentration in raw and treated groundwater sources. Dugan et al.(Dugan et al., 1995) and Koechling et al.(Koechling et al., 1997) proposed saturation type models (i.e. Monod) in which the model parameters were empirically related to the initial chlorine concentration and a measure of the organic carbon concentration. Jadas-Hecart et al.(Jadas-Hecart et al., 1992) used an immediate chlorine demand model for the first 4 hours, followed by a second-order chlorine demand model. The Water Treatment Plant model(USEPA, 1992) used three models in sequence to describe chlorine decay: 1) zero-order from $t = 0$ to 5 min, 2) second-order from $t = 5$ min to 5 hr, and 3) first-order for $t > 5$ hr.

Other research has focused on the relatively slow chlorine decay within the distribution system, assuming rapid decay reactions are complete. Wable et al.(Wable et al., 1991), Biswas et al.(Biswas et al., 1993), and Rossman et al.(Rossman et al., 1994) described "bulk" chlorine decay using first-order kinetics with satisfactory results. (Bulk decay is due to homogeneous reactions with natural organic matter, and thus ignores those reactions with compounds attached to or derived from pipe materials.) Vasconcelos et al.(Vasconcelos et al., 1996) applied four different bulk decay models, initially proposed for wastewater by Haas and Karra(Haas and Karra, 1984). The four models were: 1) first-order, 2) n^{th}-order, 3) limited first-order, and 4) parallel first-order. The results suggest the benefits of using models 2-4 "were not overwhelming" when compared to the simpler first-order model. Clark(Clark) investigated a second-order overall kinetic model that accounts for the influence of both free chlorine and a fictive reactive component intended to represent the aggregate role of all natural organic compounds (with the appropriate assumptions, this model is equivalent to (Jadas-Hecart et al., 1992)). The resulting two-species model provided results as good as, or better, than the four models investigated in (Vasconcelos et al., 1996). Powell et al.(Powell et al., 2000) compared the models of Haas and Karra(Haas and Karra, 1984) and Clark(Clark), and observed similar behavior.

The mathematical description of THM formation has primarily been examined using non-linear regression analysis(Amy et al., 1987; Chowdhury et al., 1991; Harrington et al., 1992; Owen et al., 1992; USEPA, 1992; Chowdhury et al., 1993; Lyn and Taylor, 1993). Adin et al.(Adin et al., 1991) and Clark(Clark) proposed a fictitious reaction to describe THM formation. Adin et al.(Adin et al., 1991) suggested THM formation occurs in a step-wise fashion: fast formation of a chlorinated intermediate, followed by slow formation of a second intermediate, followed by THM formation. Clark(Clark) assumed THMs form in proportion to the loss of free chlorine (the "chlorine demand"). This linear relationship was demonstrated experimentally by Clark(Clark) and Abd El-Shafy and Grünwald(El-Shafy and Grünwald, 2000).

4.2 Materials and Methods

This section discusses the source water characteristics, research specific procedures (bench scale treatment, demand study), and various standard methods for solution preparation and measurement techniques.

4.2.1 NOM Sources

Six different waters, listed in Table 4.1, were used in the various experiments to evaluate booster disinfection. Five waters were from a surface water source, while the remaining water was from a ground water source. The raw water from these sources was treated in one of three ways: full scale, pilot scale, or bench scale. Where possible, water already receiving full scale water treatment, yet before disinfection, was desired to produce results similar to those which may occur in the respective distribution system. One of the raw waters was pilot scale treated using conventional, preozonation with conventional, and preozonation, conventional, and biofiltration treatment(Miltner et al., 1996). When only raw water was available, the water was bench scale treated (see below) using coagulant doses similar to those of the actual treatment plant. All waters were $0.45 \mu m$ membrane filtered after treatment. The treated DOC concentrations ranged from 0.65 mg/L to 3.50 mg/L.

Table 4.1: Source water characteristics

Source	Location	Type[†]	Treatment Scale[‡]	Treated DOC (mg/L)	Label
Harsha Lake	Southwest, OH	S	B	3.50	HLW
City Lake	S. Norwalk, CT	S	F	2.15	CT
Ohio River	Cincinnati, OH	S	F	0.65	OH
Aquifer	Midland, TX	G	B	0.67	TX
Lake Mead	Las Vegas, NV	S	B	2.40	NV
Bull Run River	Portland, OR	S	B	1.40	–

[†]S: surface water; G: ground water
[‡]F: full scale; B: bench scale

4.2.2 Bench Scale Treatment

Bench scale treatment for the raw waters consisted of coagulation, flocculation, and sedimentation. The coagulant doses (alum, $Al_2(SO_4)_3 \cdot 18H_2O$) for the two bench scale treated waters were 80 mg/L for Harsha Lake and 0.40 mg/L for Las Vegas (based on pilot and full scale treatment respectively), and were rapid mixed at 1750 rpm for 1 minute and 1250 rpm for 3 minutes. The flocculation process occurred in three 4 minute intervals of decreasing mixer speeds: 150 rpm, 60 rpm, and 25 rpm. The flocculated water was allowed to settle a minimum of 8 hours, and then 2.0μm cartridge filtered, prior to 0.45μm filtration.

4.2.3 Demand Study

For some experiments, a residual was to be targeted at a specified time t^*. To achieve the target residual, the initial disinfectant concentration must be large enough to accommodate the decay that occurs from $t = 0$ to $t = t^*$. The demand study evaluates the residual at t^* resulting from several different initial disinfectant concentrations, based on ten chlorine to dissolved organic carbon ratios (Cl:DOC) [the various ratios were 0.25, 0.50, 0.75, 1.0, 1.5, 2.0, 2.5, 3.0, 3.5, and 4.0:1]. During each demand study, three or four duplicate samples, randomly chosen from the ten possible Cl:DOC values, were also evaluated. From these studies the initial dose which achieves the target residual at t^* was estimated.

4.2.4 Disinfectant Dosing Solutions

A concentrated hypochlorite dosing solution ($\approx$ 10 mg/mL) was prepared according to uniform formation conditions (UFC) procedure(Summers et al., 1996). A concentrated ammonia solution ($\approx$ 1 mg/mL) was prepared by dissolving 4.72 g/L of ammonia sulfate in laboratory clean water (tap water treated by reverse osmosis, ion exchange, and granular activated carbon). Both dosing solutions were buffered to pH 8.0 ± 0.2 with borate buffer. When dosing chloramine as a disinfectant, the appropriate ammonia concentration was added and mixed, prior to chlorine addition.

4.2.5 Disinfectant Measurement

The hypochlorite dosing solution concentration was measured iodometrically before each dose using Method 4500-Cl B(APHA et al., 1995). The concentration was determined by measuring the hypochlorite dosing solution concentration of three samples and averaging the results. Using a 0.1 mL graduated burette, the minimum detectable concentration of the hypochlorite solution is 0.06 mg/mL. Residual disinfectant concentrations (HOCl, NH_2Cl, $NHCl_2$) were measured using the DPD-FAS Method 4500-Cl F(APHA et al., 1995). When performing experiments, various duplicate samples were chosen randomly (which will be discussed later). Using a 0.1 mL graduated burette, the minimum detectable concentration of HOCl, NH_2Cl, and $NHCl_2$ in solution is 0.05 mg/L.

4.2.6 Ammonia Measurement

Both the ammonia dosing solution and residual ammonia concentrations were measured with an Orion 95-12 ammonia selective probe. The probe was connected to an Accumet Model 15 pH meter, which also measures electric potential (mV). A calibration curve must be developed to provide a correlation between $\Delta mV(= mV_{sample} - mV_{blank})$ and a known ammonia concentration. Six different ammonia standards were run: a blank, 0.01 mg/L, 0.03 mg/L, 0.1 mg/L, 1.0 mg/L, and 10 mg/L. The mV readings using the ammonia probe required approximately 10 minutes to provide a stable reading; thus, each sample was run only once due to time constraints. All non-zero standards were made by diluting a 1000 mg/L commercial ammonia stock solution with laboratory clean water.

4.2.7 Sample Storage

Before disinfection, treated water was stored in 55 gallon plastic drums or 30L Nalgene carboys in the dark at a constant temperature of 4°C for no more than two weeks (to minimize any degradation of water quality). Disinfected samples were stored head-space free in amber glass bottles in the dark at a constant temperature of 20°C. All glassware and storage containers were prepared "chlorine demand free" in accordance with the procedure for UFC(Summers et al., 1996).

4.2.8 THM Measurement

A small amount of sodium sulfite is added to the disinfected sample to reduce the remaining chlorine present in the sample to prevent additional THM formation between the time of sampling and the actual THM analysis. THMs are extracted from the aqueous samples using a Tekmar LSC2000 purge and trap system, and analyzed using a Varian 3400 gas chromatograph (GC). A calibration curve correlating the individual THMs ($CHCl_3$, $CHBrCl_2$, $CHBr_2Cl$, and $CHBr_3$) and area under the GC peak was developed using standard individual THM concentrations of 5 $\mu g/L$, 10 $\mu g/L$, 25 $\mu g/L$, and 50 $\mu g/L$. The calibration curve was developed before any experimental THM samples were measured, and checked after running the instrument for 8 hours. When performing experiments, various duplicate THM samples were chosen randomly (which will be discussed later).

4.2.9 DOC Measurement

Samples of treated and 0.45μm filtered water were prepared for measurement by acidifying the sample to pH 2 with concentrated phosphoric acid, and sparging with nitrogen for 5 minutes to remove any CO_2. Once prepared, the sample is analyzed for DOC using the persulfate ultraviolet oxidation, Method 5310 C(APHA et al., 1995) by a Dohrman DC-180 instrument. All samples measured for DOC were analyzed in duplicate. A 10 mg/L standard of potassium hydrogen phthalate is used to perform a one point calibration of the DOC instrument.

4.2.10 pH Measurement

pH measurements were made with a temperature correcting pH probe, Corning G-P Combo w/RJ (GP = General Purpose, RJ = Replaceable Junction; Cat. #476086), connected to a Corning pH meter, model 245. The pH meter is calibrated using a two-point calibration curve of commercially purchased pH 7.00 and pH 10.00 buffer.

4.3 Chlorine Decay - Phase I

In this section we evaluate the effect of various boost dose concentrations and times on observed chlorine decay. This work was not, however, intended to evaluate the potential for reductions in disinfectant mass or THM concentration, which may occur under booster disinfection.

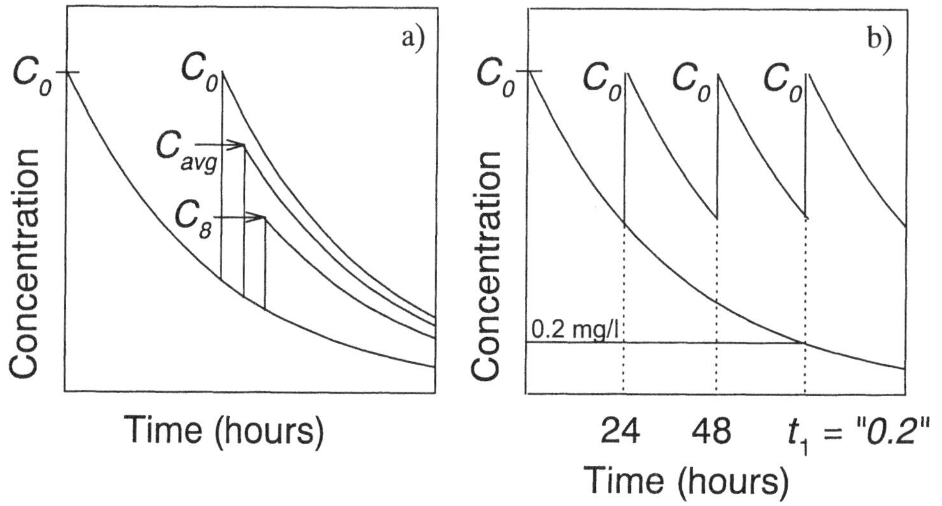

Figure 4.5: Illustration of the experimental design for a) Phase I.a, and b) Phase I.b.

4.3.1 Experimental Design

Two different sets of experiments were undertaken using laboratory scale treated Harsha Lake water (DOC = 3.5 mg/L). These Phase I experiments investigated a booster disinfection scenario with an initial dose B_0, based on a particular set of Cl:DOC, followed by: a) a single boost B_1 at time t_1 to various concentrations C_i, or b) multiple boost doses B_i, $i = 1, 2, 3$, at times t_i to initial concentration C_0.

Phase I.a

This experiment evaluated the effect of different rechlorination doses and times on disinfectant decay kinetics. Figure 4.5a illustrates the experimental design for Phase I.a. An initial dose B_0 to a concentration C_0, based on a Cl:DOC of 2.5:1, 2.0:1, or 1.5:1, was followed by a single rechlorination dose B_1 to one of three different concentrations: the initial concentration C_0, the concentration eight hours after the initial dose C_8, and the average of C_0 and C_8, C_{avg} (resulting in target dosing levels which differed by at least 1 mg/L). For each of the three rechlorination doses, three distinct rechlorination times were considered: 24 hours, 48 hours, or the final boost (which occurred when the disinfectant concentration from B_0 was approximately equal to 0.2 mg/L).

Phase I.b

This experiment evaluated the effect of sequential rechlorination of the same water on disinfectant decay kinetics. Figure 4.5b illustrates the experimental design. An initial dose B_0 to a concentration C_0, based on a Cl:DOC of 2:1, 1.5:1, or 1:1, was followed by sequential

Table 4.2: Boost concentration for Phase Ia chlorine studies

Target Boost Conc. (mg/L)	Cl:DOC		
	2.5:1	**2.0:1**	**1.5:1**
C_0	8.75	7.00	5.25
C_{avg}	6.79	5.63	3.87
C_8	4.83	4.26	2.50

rechlorination doses B_i to C_0 at 24 hours, 48 hours, and after the concentration reached a residual of 0.2 mg/L, or 15 days after the 48 hr boost, which ever occurred first.

4.3.2 Results and Discussion: Phase I.a

Figure 4.6 shows the disinfectant decay data for the experiments with a Cl:DOC of 2.5:1 (representative of the results with a Cl:DOC of 2.0 and 1.5:1 in Appendix C). The solid symbols represent the chlorine concentration of the initial dose, while the hollow symbols represent the chlorine concentration from a single boost dose to one of the three concentrations and times. The three boost levels for the different experiments are given in Table 4.2. There is, however, no data presented for the 48 hour boost doses based on a Cl:DOC of 2:1; the DPD indicator solution used in the chlorine residual measurement was prepared improperly causing large measurement errors.

The effect of disinfectant concentration, after the boost dose, on disinfectant decay is difficult to see from Figure 4.6. Instead, Figure 4.7 shows the respective chlorine demand for the experimental data with a Cl:DOC of 2.5:1, where chlorine demand is defined as the chlorine concentration consumed by reaction at time t (representative of the results with a Cl:DOC of 2.0 and 1.5:1 in Appendix C). By plotting chlorine demand, the effect of the boost dose concentration is more apparent since the chlorine demand at the time of a boost is the same regardless of the boost dose magnitude. For all of the Cl:DOC and boost times, increasing the concentration resulting from a boost increased the disinfectant decay rate, as evidenced by the increased chlorine demand. The observed relationship between disinfectant decay rate and boost concentration also holds for the initial doses. Figure 4.8 shows increasing the initial Cl:DOC increases the chlorine demand at each of the three boost times. This indicates, for this particular water and experimental conditions, the disinfectant demand at a particular time (and, therefore, the disinfectant decay rate) is dependent on chlorine concentration.

The effect of boost time on disinfectant decay can also be evaluated from these experiments. It should be noted, however, that time does not really affect chlorine decay; it is simply an easy surrogate for discussing how far the associated reaction(s) may have progressed. Another, potentially more accurate, representation of how far the reactions have

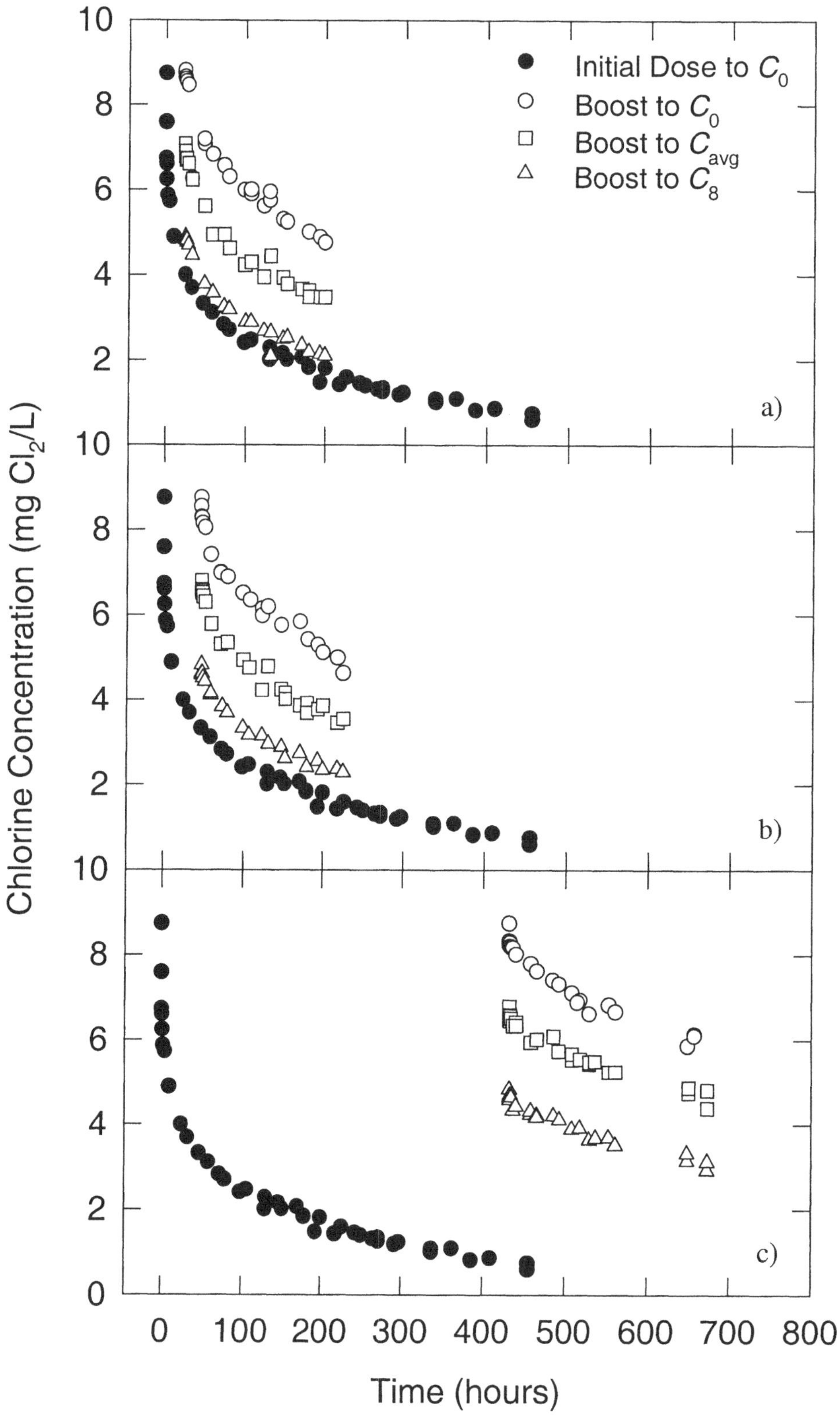

Figure 4.6: Chlorine decay data of Phase Ia experiments based on initial Cl:DOC of 2.5:1 with boost doses at: a) 24 hrs., b) 48 hrs., and c) when residual of initial dose is 0.2 mg/L.

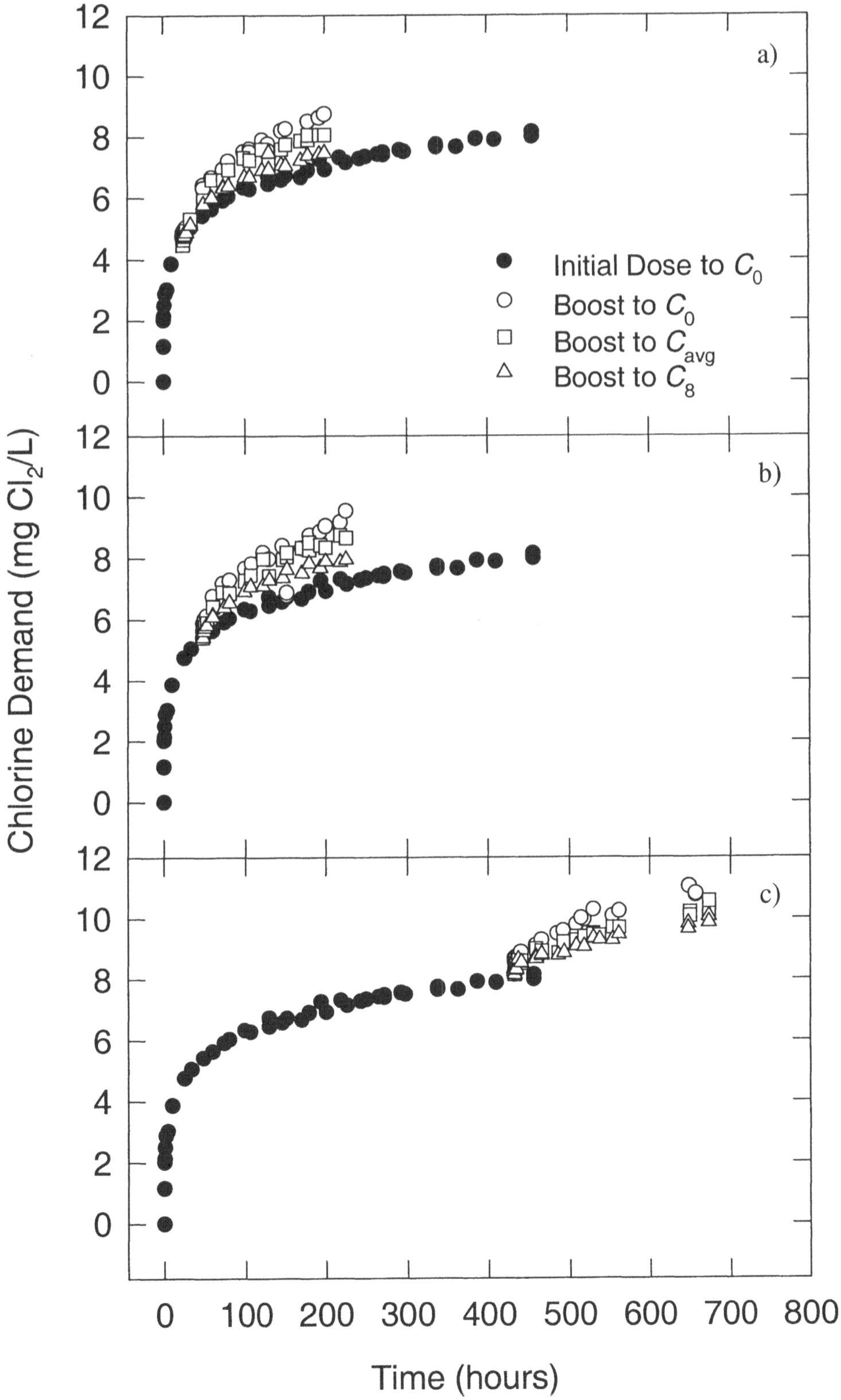

Figure 4.7: Chlorine demand data of Phase Ia based on an initial Cl:DOC of 2.5:1 with boost doses at: a) 24 hrs., b) 48 hrs., and c) when the residual from the initial dose is 0.2 mg/L.

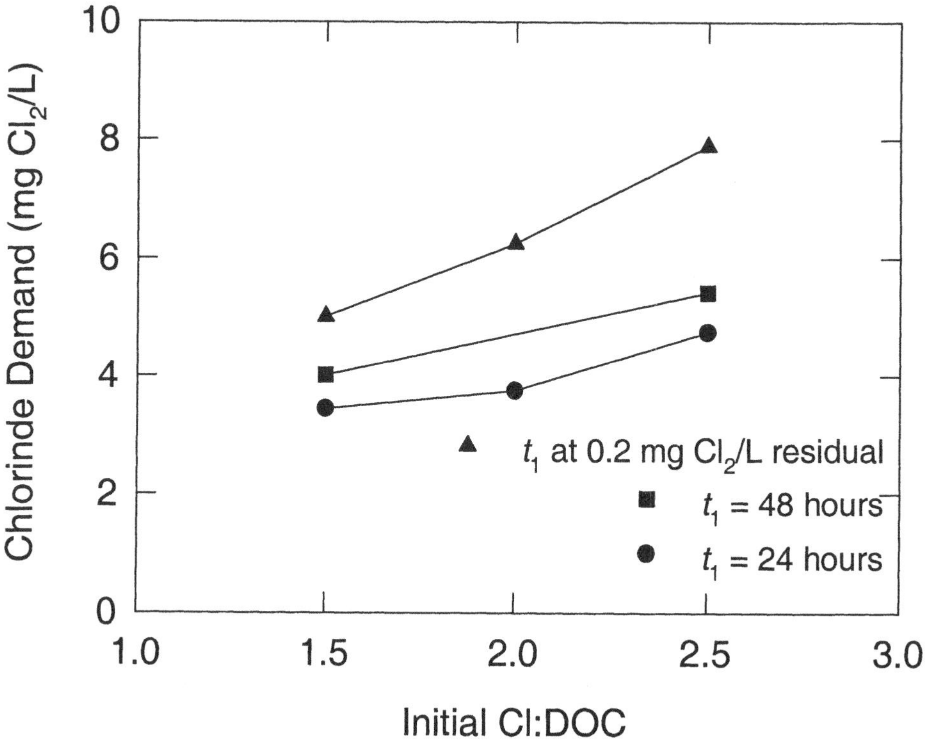

Figure 4.8: Chlorine demand prior to boosting resulting from different initial chlorine doses.

progressed is by observing the chlorine demand. An evaluation of chlorine decay characteristics after boosting to a particular chlorine concentration – at different levels of chlorine demand – may provide insight into the effect of reaction time. Figure 4.9 shows the chlorine demand after boosting to a single concentration (either C_0, C_{avg}, or C_8 at $t_1 = 24$, 48, and when residual falls to 0.2 mg/L) for Cl:DOC of 2.5:1 (representative of the results for Cl:DOC of 2.0 and 1.5:1 in Appendix C).

These results show that the the chlorine demand decreases with an increase in the pre-boost chlorine demand C_d (NOTE: The symbol shapes in Figure 4.9 corresponds to the different boost levels in Figures 4.6 and 4.7, while the symbol fill type represents the different boost times and chlorine demand at t_1; e.g., the data in Figure 4.9a correspond to the C_0 data in Figure 4.6a, b, and c). The influence of previous chlorine demand on the disinfectant decay is evident in the demand which occurs after the 24 and 48 hour boost doses (Figure 4.9). Boosting at 48 hours requires more chlorine mass than boosting at 24 hours to provide similar chlorine concentrations, due to the additional decay. One might expect the demand to be larger for the system with more applied chlorine, however, the associated demand following the two boosts behave similarly. Therefore, the expected increase in decay after the 48 hour boost, due to the increased application of chlorine, is offset by a smaller amount of oxidizable material, causing the demand after the two boosts to behave similarly. Disinfectant decay, under booster disinfection, is both a function of the boost dose concentration and the chlorine demand at the time of boosting.

4.3.3 Results and Discussion: Phase I.b

Figure 4.10 illustrates the disinfectant decay observed with multiple boost doses, each to provide a concentration equivalent to the initial concentration, for three different Cl:DOC. Figure 4.11 illustrates the changing chlorine demand with the additional boosts.

The increase in chlorine demand after the boost doses at 24 and 48 hours (Figure 4.11) is obviously larger than the increase after the final boost. To show this more clearly, figure 4.12 illustrates the chlorine demand after each booster dose in the multiple dose experiment; each dose is applied to water which has a different chlorination history, and thus a different pre-boost chlorine demand. These results show the chlorine demand decreases with successive boost doses, even though the concentrations immediately after boosting are identical. For the data shown in Figure 4.12 (and those in Figure 4.9), the last boost dose requires the largest mass of chlorine to achieve the desired boost concentration (for Figure 4.12, the "mass of chlorine" is the cumulative mass added, not the amount for the last boost); the last boost also, however, is associated with the least disinfectant decay. These results reinforce that the decay after booster disinfection is a function of both the chlorine concentration and previous chlorine history (as measured, for example, by the cumulative chlorine demand).

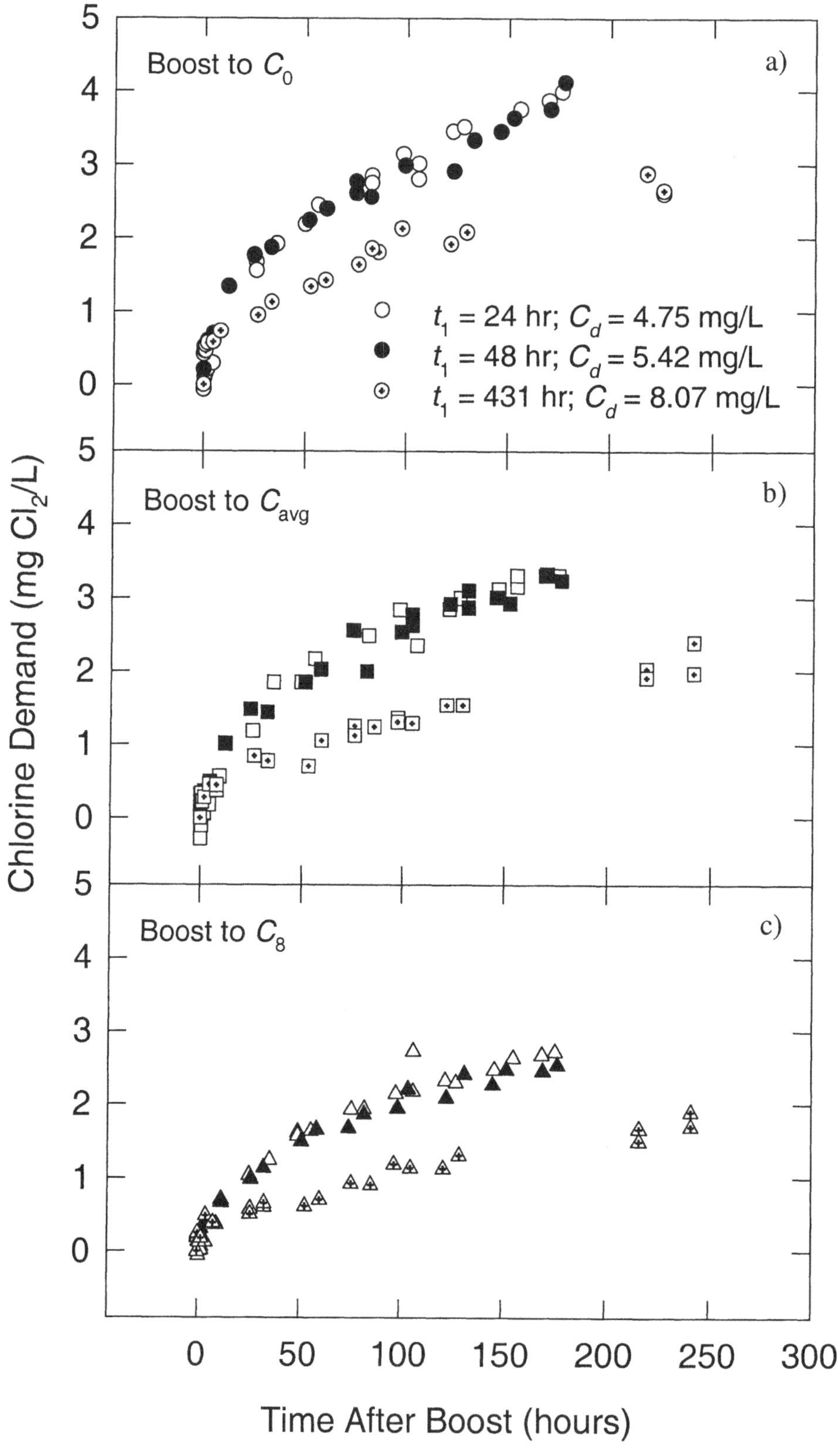

Figure 4.9: Chlorine demand data following one of three boost doses of Phase Ia; Cl:DOC = 2.5:1.

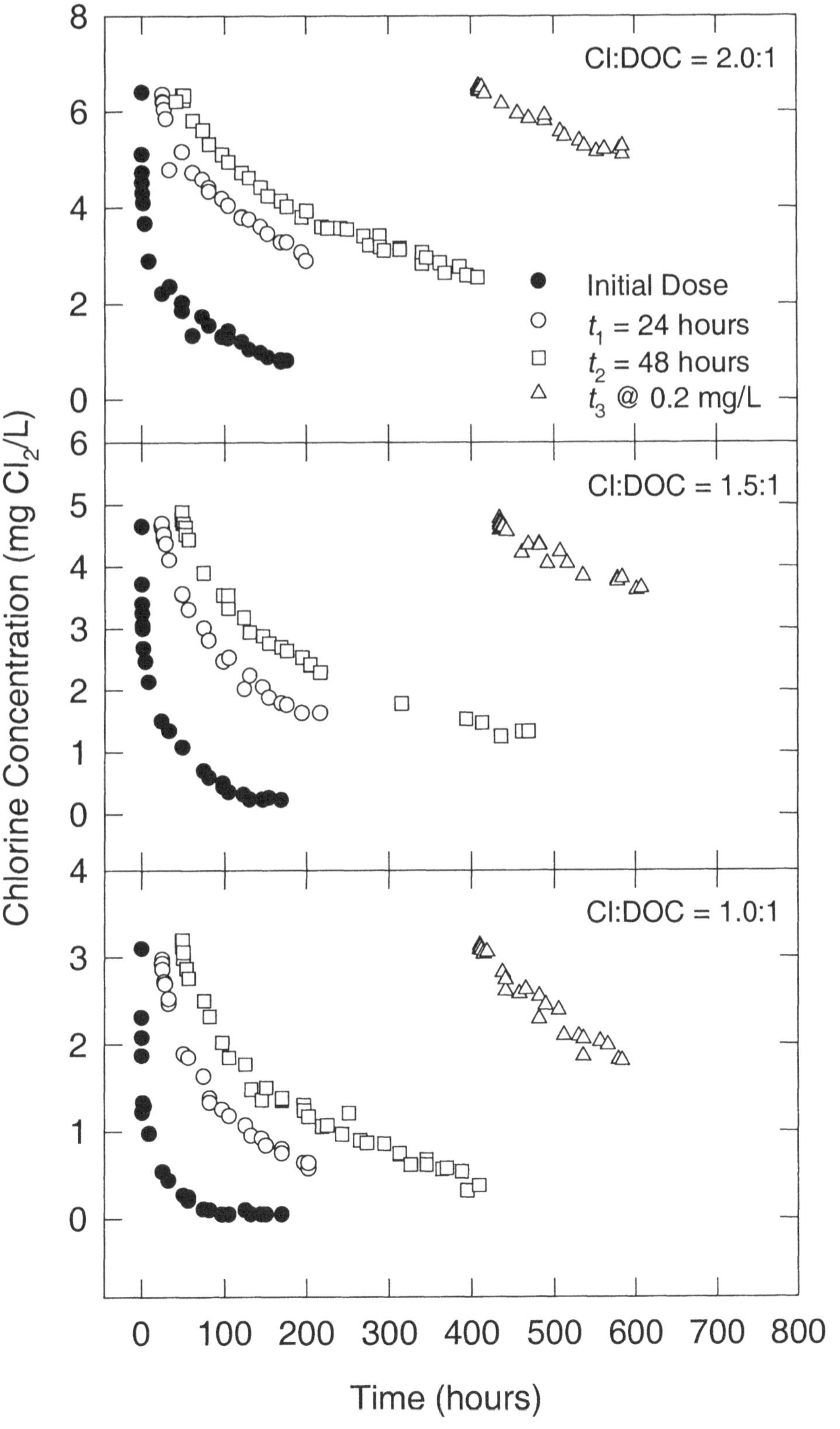

Figure 4.10: Chlorine decay for multiple boost doses of Phase Ib; initial Cl:DOC = 2:1, 1.5:1, and 1:1.

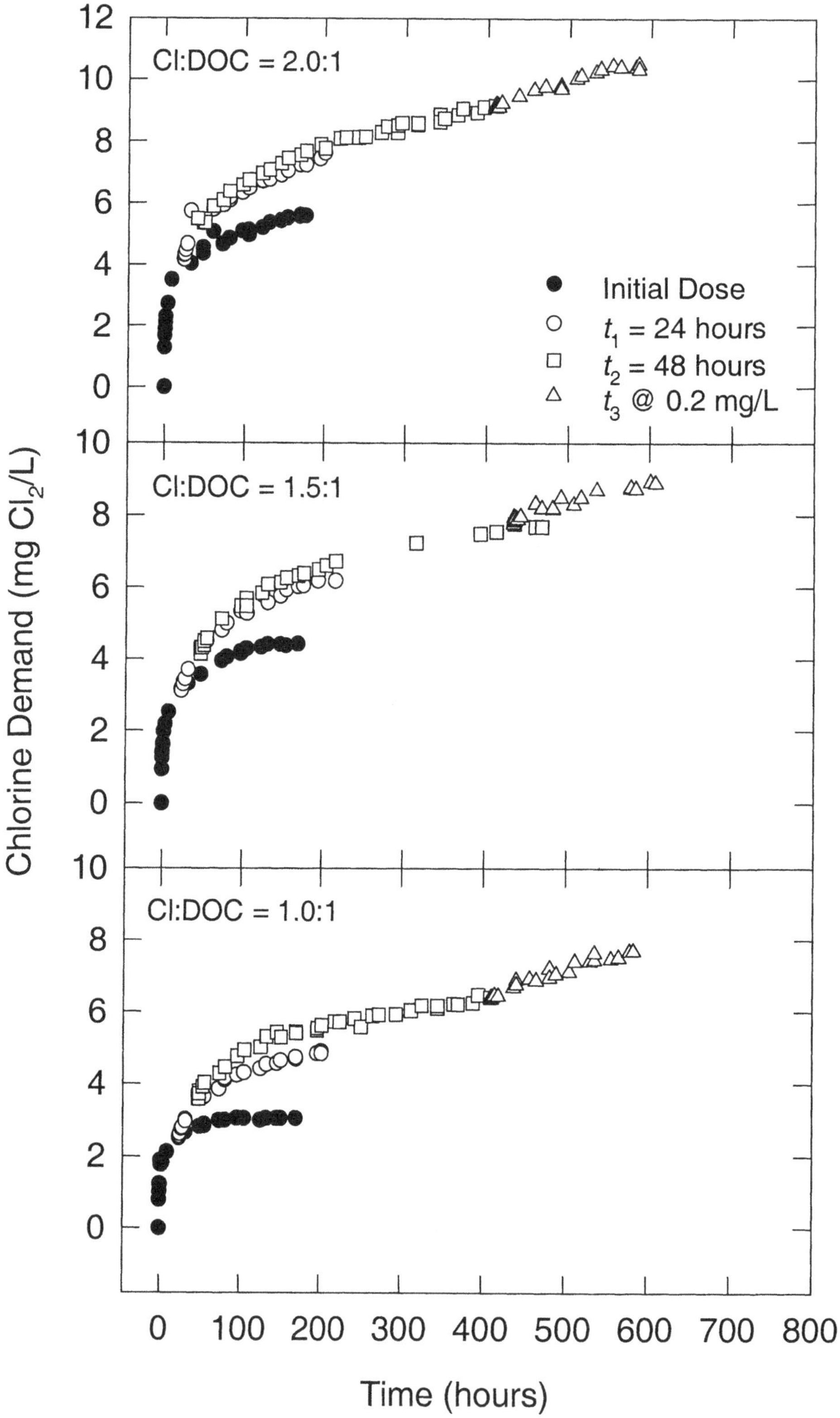

Figure 4.11: Chlorine demand for multiple boost doses of Phase Ib; initial Cl:DOC = 2:1, 1.5:1, and 1:1.

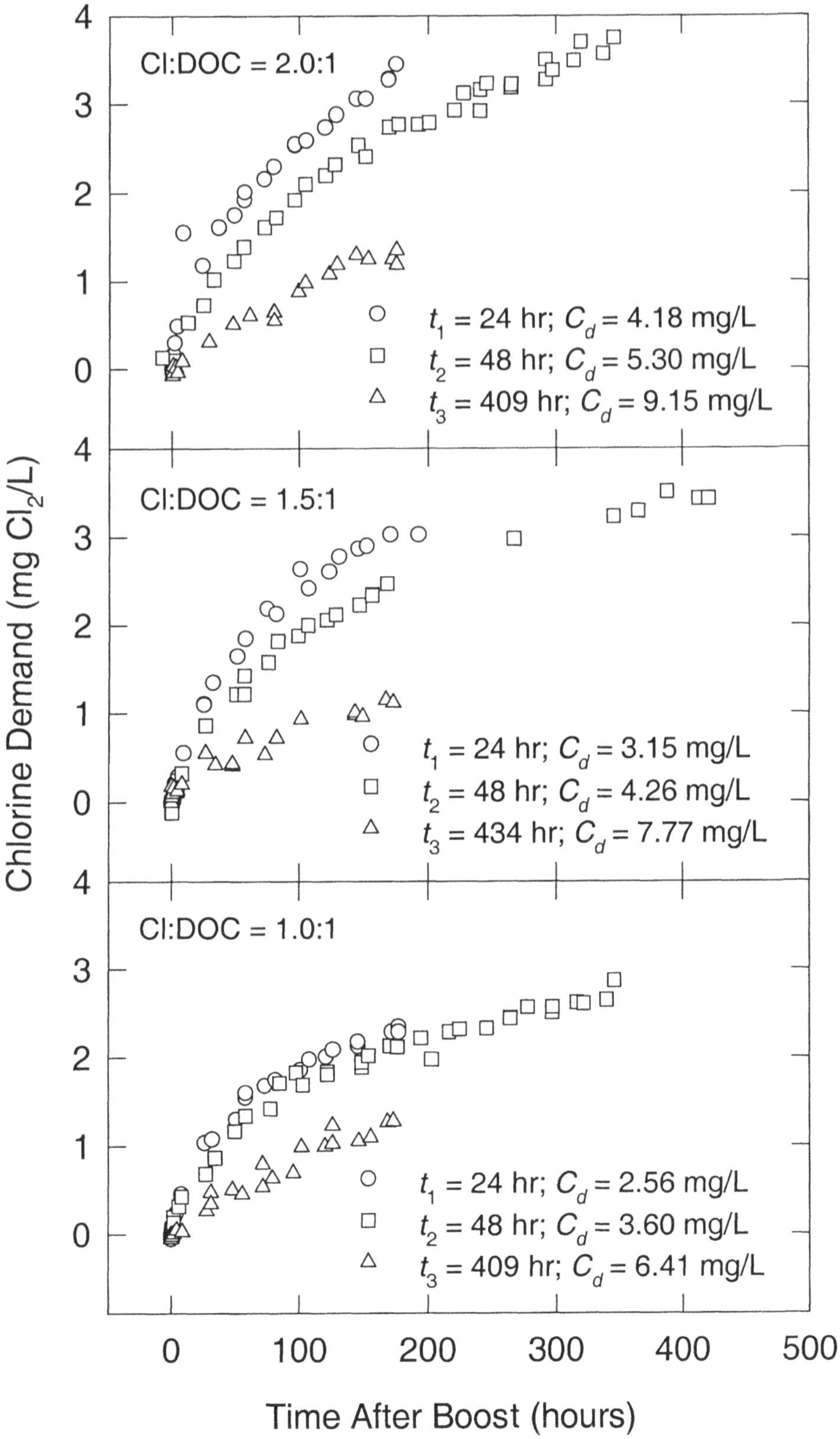

Figure 4.12: Chlorine demand after each boost dose of Phase Ib; initial Cl:DOC = 2:1, 1.5:1, and 1:1.

4.3.4 Summary

The qualitative effects of boost dose and time on disinfectant decay were similar regardless of the number of boost doses applied. For a particular boost time, including the initial dose, increasing the chlorine concentration increased the chlorine decay. Delaying the time of a boost dose, when boosting to a particular concentration, decreases the chlorine decay after the boost; this is explained by the decreased amount of oxidizable organic material that remains. The use of chlorine demand as a reactive component surrogate may better explain chlorine decay results, especially under booster conditions where a water is likely to experience the same chlorine residual after different chlorination histories. An accurate description of chlorine decay after a boost dose will likely require knowledge of its chlorination history, at least as measured by the pre-boost chlorine demand.

4.4 Chlorine Decay and THM Formation - Phase II

In this section we evaluate chlorine decay and THM formation under various booster disinfection scenarios. The main objective is to assess the potential benefits of booster disinfection to reduce the applied disinfectant mass yet still maintain an adequate disinfectant residual, and similarly to reduce the associated THM formation.

4.4.1 Experimental Design

This set of experiments evaluated the potential of booster chlorination to reduce the amount of chlorine applied compared to conventional chlorination, while maintaining an adequate free chlorine residual (adequate residual is defined as 0.2 mg/L of free chlorine for this study). A conventional chlorination scenario was compared to an *equivalent mass* booster chlorination scenario such that $M_0 = \sum_{i=0}^{n_b} B_i$, where M_0 is the conventional dose at time t_0, B_i is the boost dose at time t_i, and n_b is the number of boost doses after the initial dose B_0. If booster chlorination can reduce the average disinfectant decay then, under this equivalent mass scenario, booster chlorination should produce a higher residual after an equivalent mass is added. Figure 4.13 illustrates the excess residual that may result under a mass equivalent booster chlorination scenario. Given this result, one of the booster scenario doses could be reduced while still maintaining the desired residual at time t_i. For phase II, the initial chlorine doses, M_0 and B_0, were based on: a) a predetermined set of Cl:DOC ratios and b) a Cl:DOC that provide a 0.2 mg/L residual at the time of boost, or at the end of the experiment; this Cl:DOC was determined by a separate chlorine demand study.

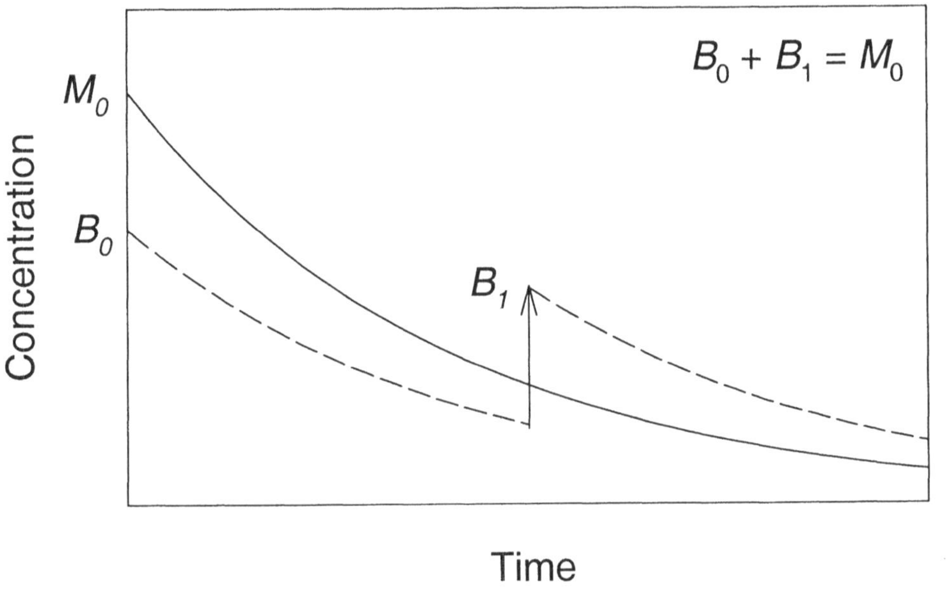

Figure 4.13: Theoretical advantage of booster disinfection if disinfectant decay is a function of chlorine concentration.

Phase II.a

For the pilot scale treated Harsha Lake water and full scale treated South Norwalk water, the chlorine decay of a single conventional dose M_0 was compared to two or three different booster chlorination scenarios. Figure 4.14 illustrates the single dose booster chlorination scenario for the Harsha Lake water experiments. M_0 was determined by a predetermined Cl:DOC and B_0 was set at 50% or 75% of M_0. As mentioned earlier, the Harsha Lake water was sampled at three points in the pilot plant. The DOC of the different treatment options were 2.7 mg/L for conventional treatment, 2.6 mg/L for conventional treatment followed by ozonation, and 2.0 mg/L for conventional, ozonation, and biofiltration treatment. Ozonation and biofiltration have been shown to decrease chlorine demand(Koechling et al., 1997; Matheis et al., 1998). M_0 and B_0 were determined using a Cl:DOC of 1.5:1. A single boost dose B_1 was applied to the water when the residual concentration of B_0 was approximately 0.2 mg/L. B_1 applied enough chlorine to produce a mass equivalent system (e.g. $B_1 = 0.25M_0$ for the scenario with $B_0 = 0.75M_0$).

The DOC of the full scale treated South Norwalk water was 2.65 mg/L. Similar sets of experimental runs were made with M_0 determined by a Cl:DOC of 2:1 and 3:1, and B_0 equal to $0.75M_0$ or $0.50M_0$, as shown in Figure 4.15. Unlike the Harsha Lake water experiment, the application of boost doses for this experiment were based on time, either 24 or 72 hours. These experiments additionally evaluated a booster scenario with two sequential boosts as well as a single boost. For $B_0 = 0.50M_0$, a single dose B_1 applied a mass equivalent dose

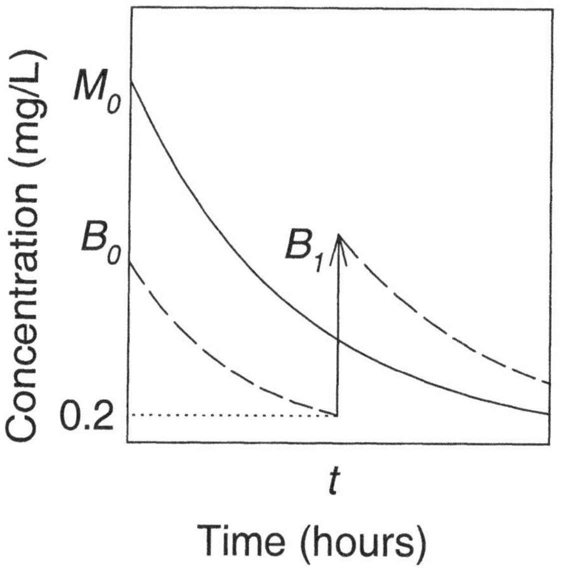

Figure 4.14: Illustration of the experimental design for the single boost chlorination scenario of Phase II.a with Harsha Lake water.

at $t = 24$ hours or $t = 72$ hours or, for the two dose experiment, $B_1 = B_2 = (M_0 - B_0)/2$, which were applied at $t = 24$ hr and 72 hr, respectively.

Phase II.b

Disinfectant decay and THM formation under booster chlorination were evaluated for the full scale treated South Norwalk and Cincinnati waters, and bench scale treated Midland and Las Vegas waters. With few exceptions, the experimental design, shown in Figure 4.16, was identical for all waters. M_0 was based on a Cl:DOC determined by a demand study to maintain a 0.2 mg/L residual after 120 hours (the length of the experiment). B_0 was based on a Cl:DOC determined by a demand study to maintain a 0.2 mg/L residual after 24 or 72 hours (the time of boost doses). These experiments also evaluated a booster scenario with two sequential boosts as well as a single boost. For the single dose experiment, B_1 applied a mass equivalent dose at $t = 24$ hr or $t = 72$ hr. For the two dose experiment (done for all of waters except Las Vegas), $B_1 = B_2 = (M_0 - B_0)/2$, which were applied at $t = 24$ hr and 72 hr respectively. For the Las Vegas water, the single booster dose scenarios were also repeated at a pH of 9.8 (in addition to UFC pH = 8.0).

As with the Phase I experiments, the data collected is presented as both chlorine concentration and chlorine demand against time. Additionally, for Phase IIb, THM formation and demand study data are presented; these may provide an estimate of the kinetic benefit (reduction of mass needed to satisfy residual) due to booster disinfection, for a particular water and dose scenario (as shown, this benefit is not guaranteed). The experimental data

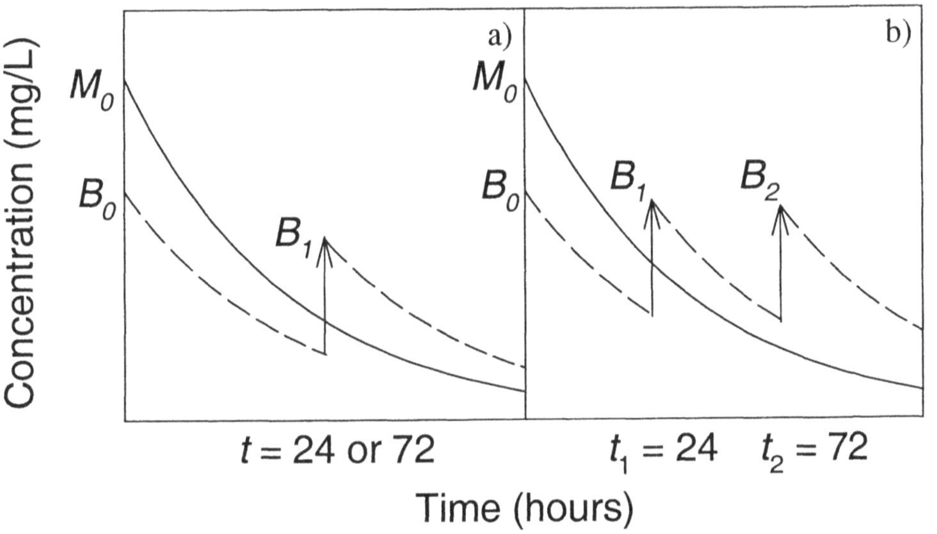

Figure 4.15: Illustration of the experimental design for a) single and b) multiple boost chlorination scenario of Phase II.a with the South Norwalk, CT water.

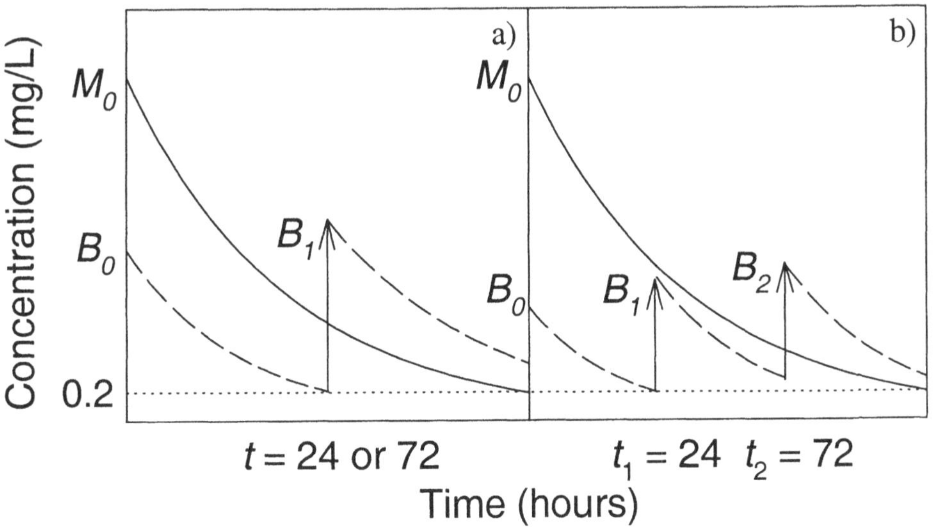

Figure 4.16: Illustration of the experimental design for a) single and b) multiple boost chlorination scenario of Phase II.b.

Table 4.3: Boost dose times (hrs.) for phase IIa Harsha Lake chlorine studies.

Treatment	Boost times (hrs.)	
	$B_0 = 0.75M_0$	$B_0 = 0.50M_0$
Conventional	26.5	5.4
Conventional/Ozonation	41.1	17.1
Conv./Ozone/Biofiltration	46.1	8.9

presented in the various figures are identified by solid symbols for conventional disinfection scenarios, and hollow symbols for booster disinfection scenarios.

4.4.2 Results and Discussion: Phase II.a

Harsha Lake

The approximate times that the boosts occurred during the Harsha Lake experiments are shown in Table 4.3 for the different treatment options. Figures 4.17, 4.18, and 4.19 show the disinfectant decay for the differently treated water and disinfectant dosing scenarios at a Cl:DOC of 1.5:1 for the total dose. None yield a residual at the end of the run higher than the M_0 dose, thus the equivalent mass scenarios would not allow either the initial or boost dose to be reduced and still maintain a 0.2 mg/L residual. Figure 4.18.b does show an advantage, however, this scenario was not mass equivalent; the boost dose was approximately 0.5 mg/L too large. Also, the difference in disinfectant residual between the conventional and booster scenarios is less than the excess chlorine applied to the system, which may suggest that even if the system were mass equivalent an advantage may not have been realized due to the increased decay after boosting.

Figure 4.20 shows the chlorine demand over time for the concentration data in Figure 4.19 (these data are representative of chlorine demand associated with Figures 4.17 and 4.18, see Appendix C) . The demand of M_0 and B_0 at $t \leq t_b$ is very similar indicating that disinfectant decay during this time and at these concentrations is essentially independent of chlorine concentration. Therefore, after boosting to achieve mass equivalency, the resulting booster concentration is essentially the same as the conventional scenario and with a similar chlorine demand. A chemical kinetic advantage is not realized.

South Norwalk

Given the similarity in results, only the data for $B_0 = 0.50M_0$ will be discussed. The experimental data for $B_0 = 0.75M_0$ is presented in Appendix C. The corresponding discussion, however, is relevant to both initial boost dose scenarios.

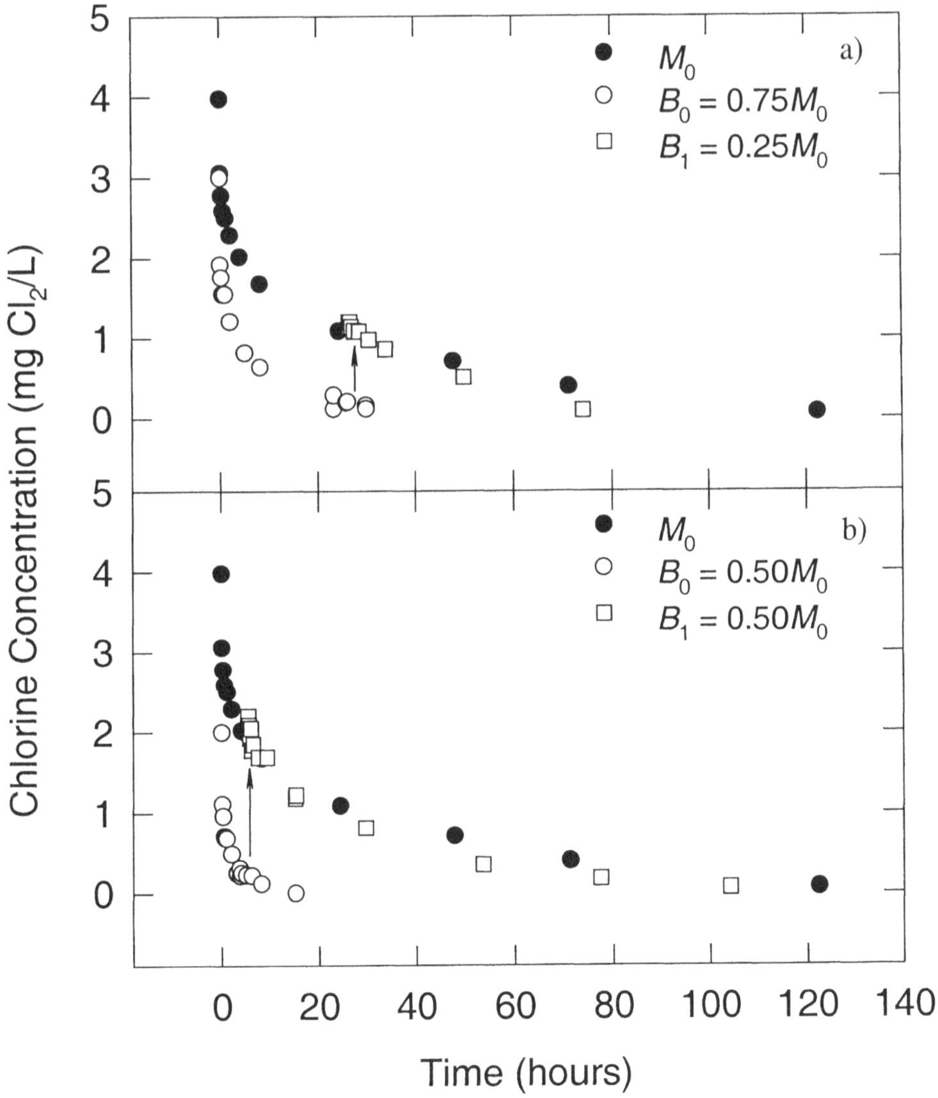

Figure 4.17: Chlorine decay data of phase IIa for conventionally treated Harsha Lake water with B_0 determined by a) $0.75M_0$ and b) $0.5M_0$; Cl:DOC = 1.5:1.

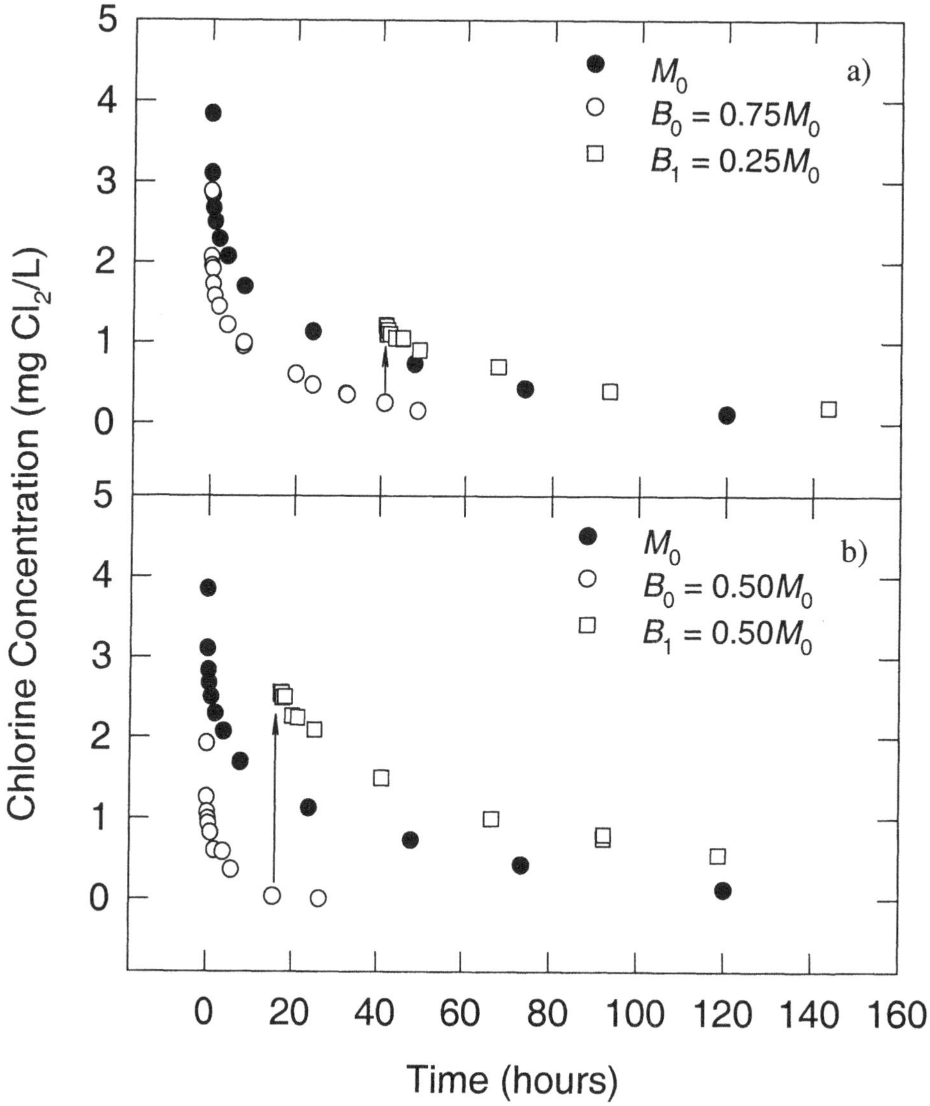

Figure 4.18: Chlorine decay data of phase IIa for conventional/ozone treatment of Harsha Lake water with B_0 determined by a) $0.75M_0$ and b) $0.5M_0$; Cl:DOC = 1.5:1.

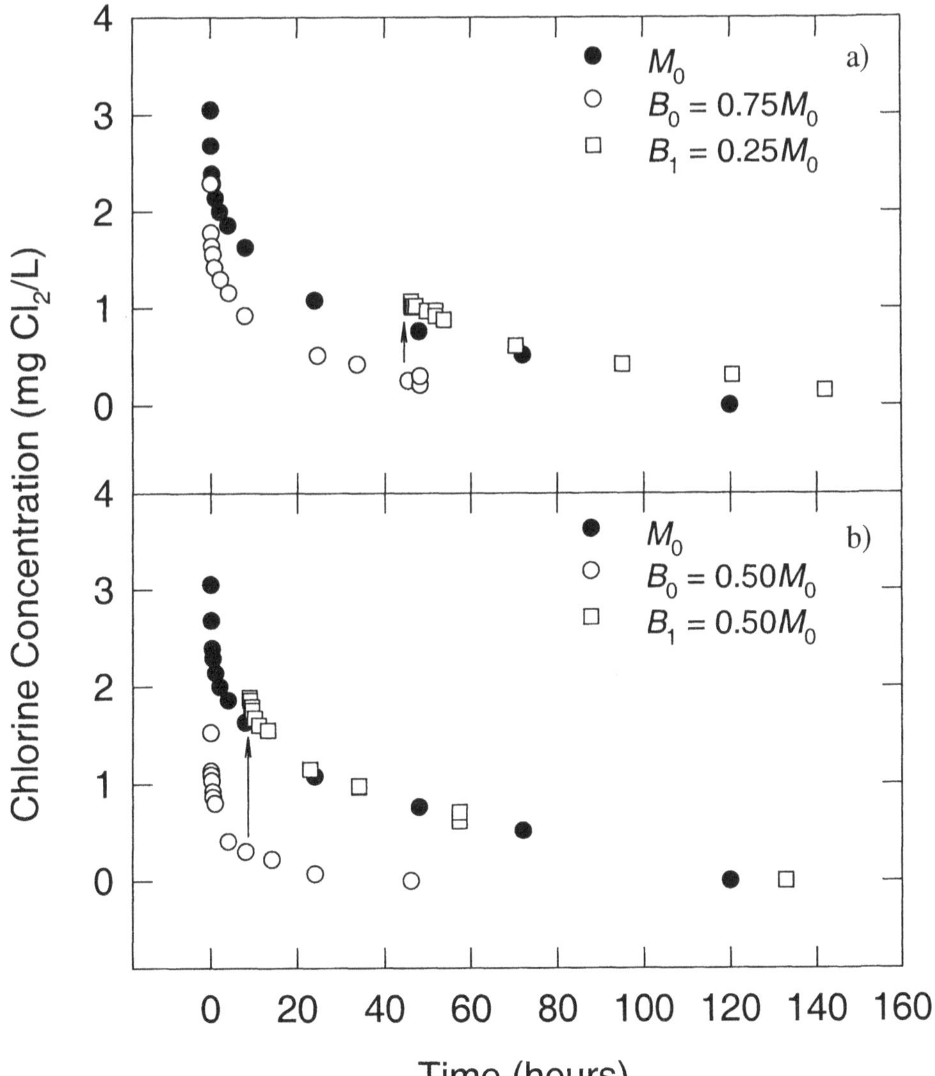

Figure 4.19: Chlorine decay data of phase IIa for conventional/ozone/biofilter treatment of Harsha Lake water with B_0 determined by a) $0.75M_0$ and b) $0.5M_0$; Cl:DOC = 1.5:1.

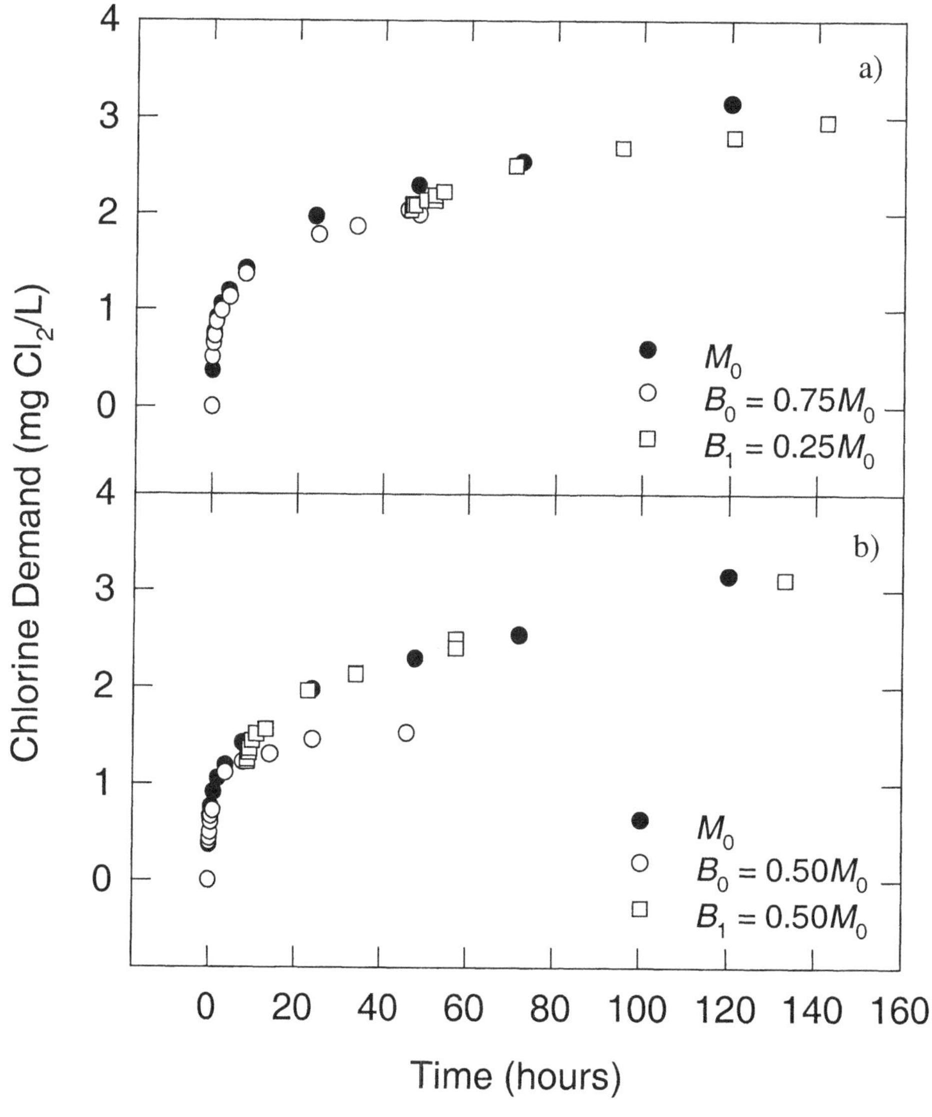

Figure 4.20: Chlorine demand data of phase IIa for conventional/ozone/biofilter treatment of Harsha Lake water with B_0 determined by a) $0.75M_0$ and b) $0.5M_0$.

For a Cl:DOC of 3:1, Figure 4.21 shows no chemical advantage for residual maintenance for any of the booster scenarios. For a Cl:DOC of 2:1, however, Figure 4.22 illustrates booster disinfection produces a larger residual immediately after boosting. This difference in booster and conventional residual decreases as the reaction proceeds becoming almost non-existent for the single boost scenario occurring at 24 hours (Figure 4.22.a).

Evaluating chlorine demand against time, Figure 4.23, for a Cl:DOC of 3:1, shows little or no difference in chlorine demand between M_0 and B_0 at $t < t_b$. Again, this indicates the reactions under these particular conditions are independent of chlorine concentration. Figure 4.24, for a Cl:DOC of 2:1, however, shows decreased chlorine demand for B_0, with respect to M_0, for all of the different booster disinfection scenarios at $t < t_b$. This decreased chlorine demand allows the boost doses to provide a higher residual than the associated conventional dose immediately after dosing. These results reinforce the intuitive notion that chlorine must limit the reaction in order for oxidation, and thus chlorine demand, to depend on the chlorine concentration. As chlorine doses are decreased chlorine will become limiting at some water specific dose. The dose where chlorine becomes limiting also depends on the chlorination history (the pre-boost demand) and likely will decrease as chlorine demand increases (since less oxidizable organic matter remains). Thus determining the conditions under which booster chlorination will yield a kinetic advantage is not straightforward, and must be addressed experimentally for a unique water and expected dosing conditions.

In attempting to describe the disinfectant decay observed between the disinfection scenarios based on a Cl:DOC of 3:1 and 2:1, it is proposed that the chlorine demand, before boosting, is related to the initial chlorine dose by saturation type behavior e.g. Michaelis-Menten(Levenspiel, 1979). Figure 4.25 illustrates how this behavior may occur at the boost time t_b. For high chlorine doses, e.g., Cl:DOC of 3:1, the demand of M_0 and B_0 are the same at t_b, therefore, the boost dose, to achieve mass equivalency, will produce a similar concentration given the chlorine demand of the boost scenario water is similar to that of the conventional dose. For low chlorine doses, e.g., Cl:DOC of 2:1, however, the demand at t_b is a function of concentration. This decrease in demand for B_0, relative to M_0, allows a mass equivalent boost dose to provide a higher concentration, with respect to conventional disinfection, immediately following a boost. This type of curve cannot imply anything about the decay after the boost as it is a function of both concentration and demand at the time of the boost, both of which are different for the two disinfection scenarios after boosting. Given this information, it may be possible to achieve an even larger benefit from booster disinfection by shifting the doses M_0 and B_0 further to the left (Figure 4.25) to increase their difference. This concept led to the use of the demand study to determine the lowest values of M_0 and B_0, which will still produce a 0.2 mg/L residual, in an attempt to produce a greater benefit from booster disinfection. Since this minimum value of B_0 also depends on the reaction time prior to the boost dose, a reduction in residence time between boost doses (i.e. increasing the number of doses) should also enhance any kinetic benefit of booster disinfection. The experiments did not, however, investigate this possibility.

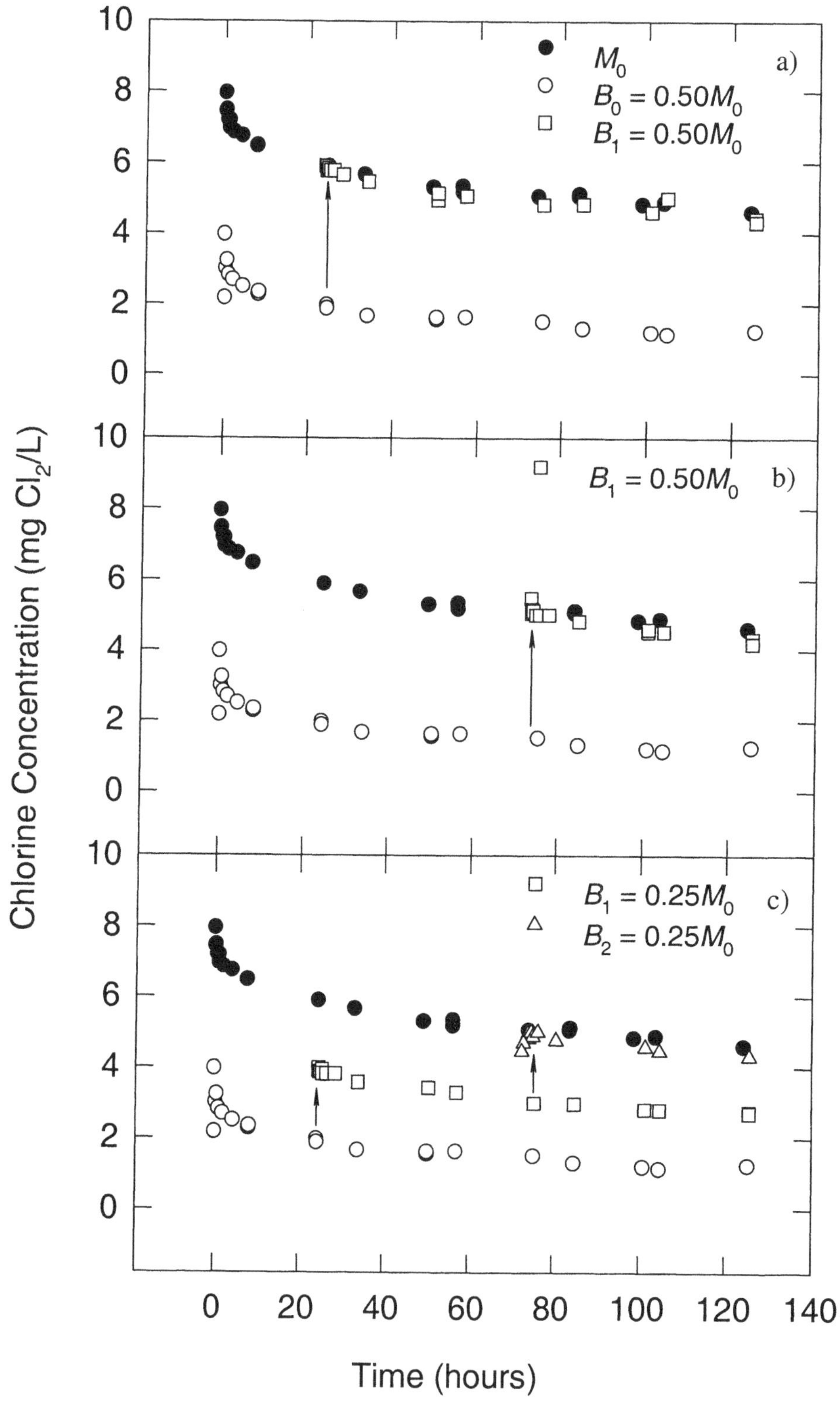

Figure 4.21: Chlorine decay data of phase IIa for South Norwalk water with boost doses at a) 24 hrs., b)72 hrs., and c) 24 and 72 hrs.; $B_0 = 0.50M_0$, Cl:DOC=3:1.

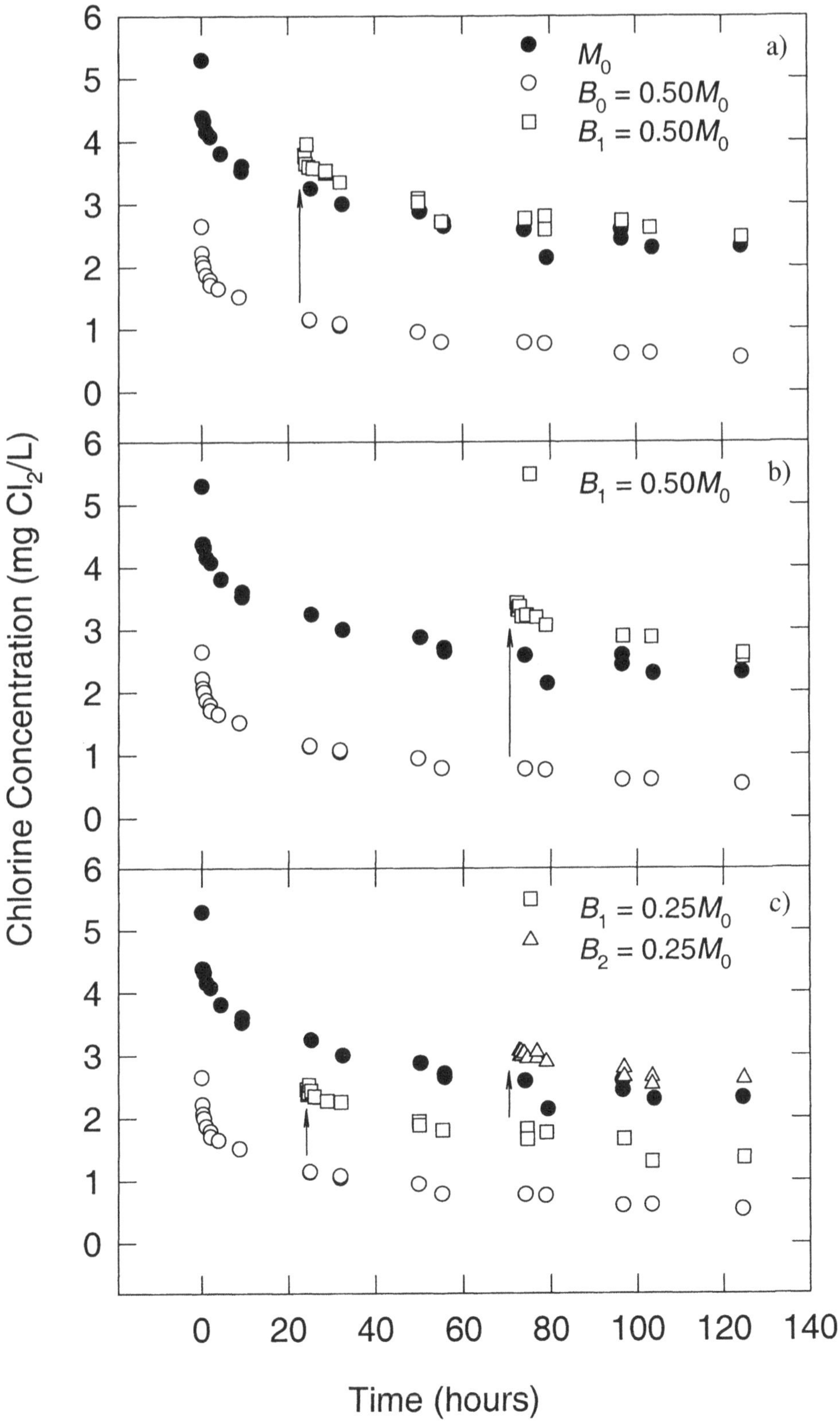

Figure 4.22: Chlorine decay data of phase IIa for South Norwalk water with boost doses at a) 24 hrs., b)72 hrs., and c) 24 and 72 hrs.; $B_0 = 0.50M_0$, Cl:DOC=2:1.

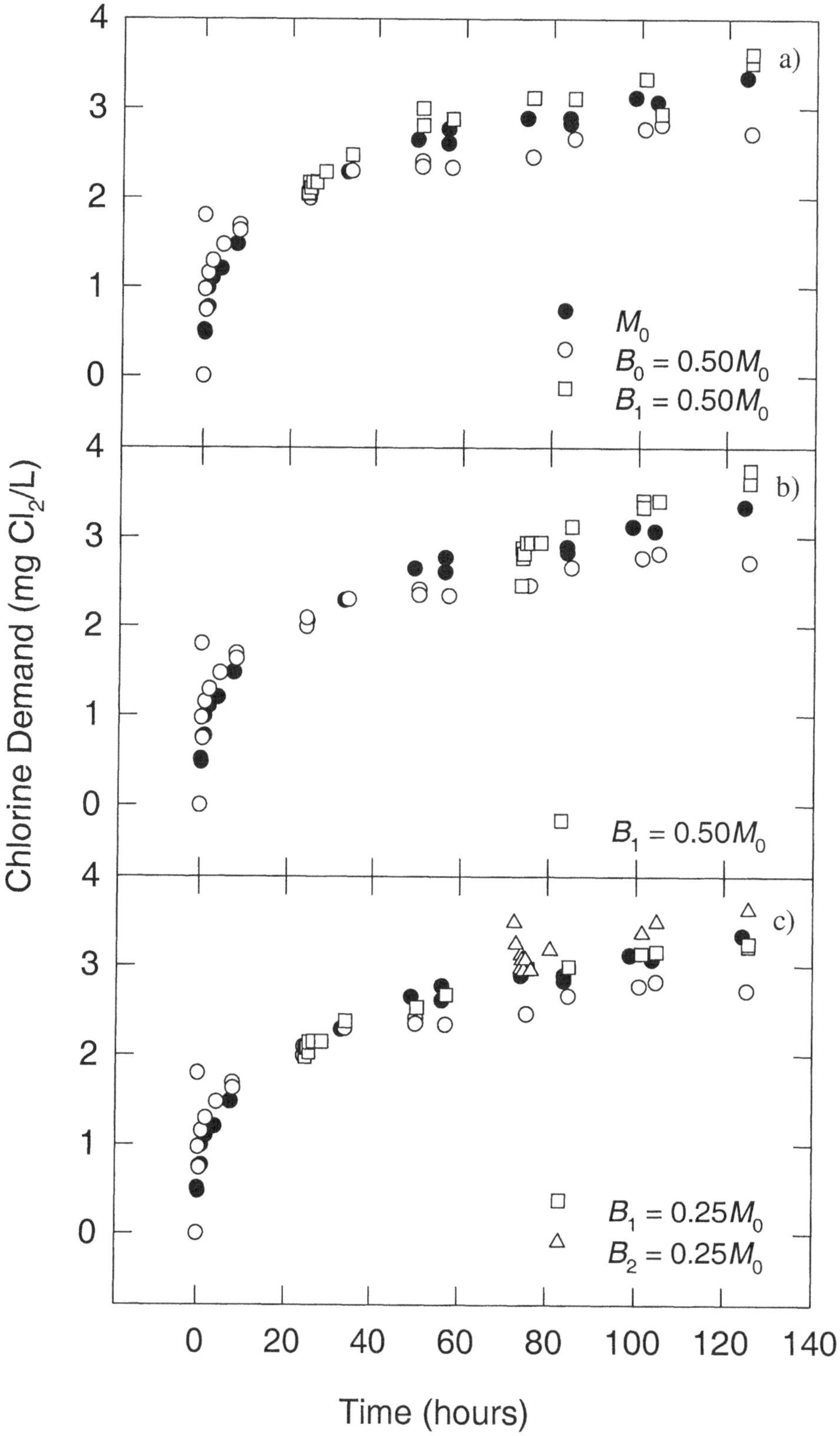

Figure 4.23: Chlorine demand data of phase IIa for South Norwalk water with boost doses at a) 24 hrs., b)72 hrs., and c) 24 and 72 hrs.; $B_0 = 0.50M_0$, Cl:DOC=3:1.

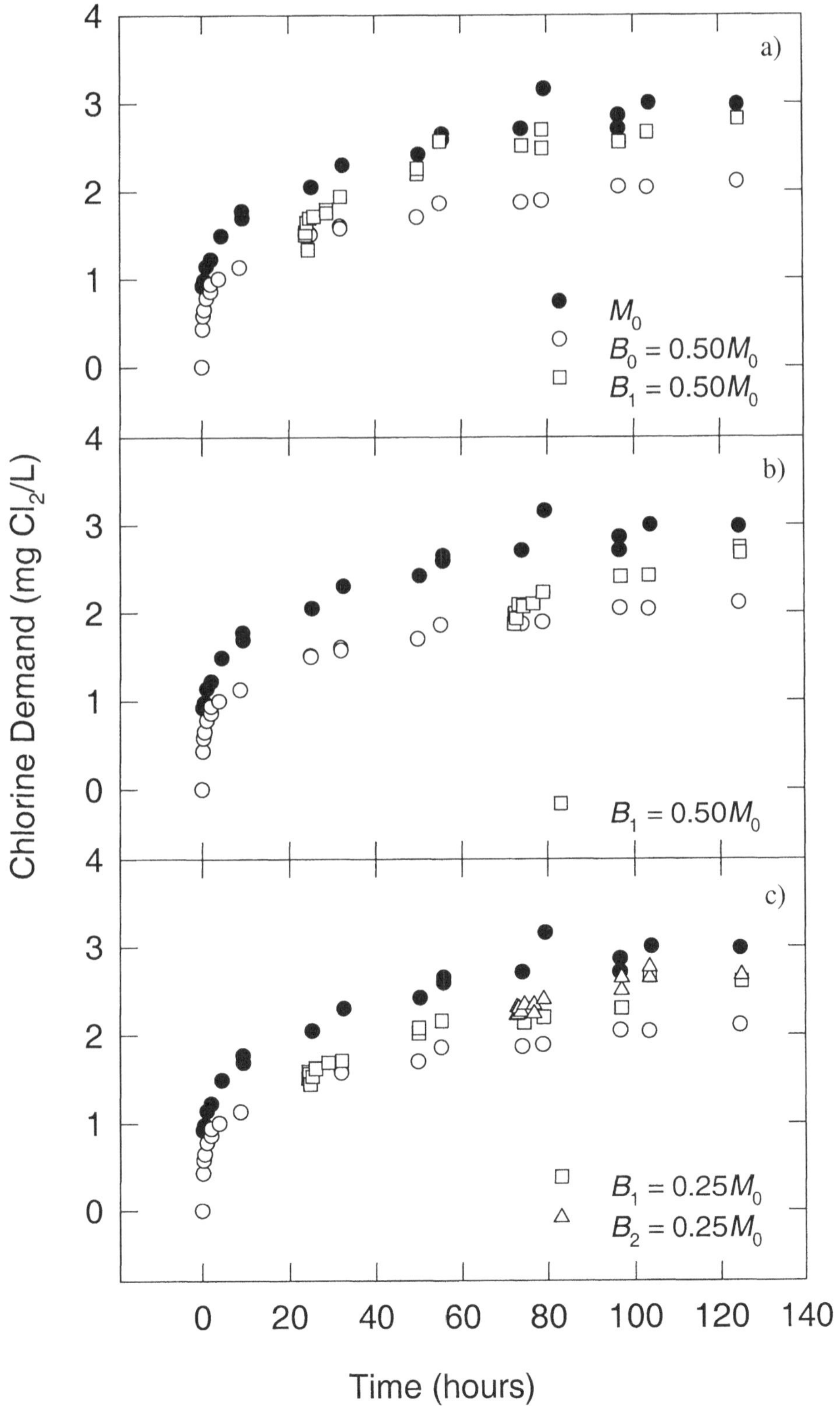

Figure 4.24: Chlorine demand data of phase IIa for South Norwalk water with boost doses at a) 24 hrs., b)72 hrs., and c) 24 and 72 hrs.; $B_0 = 0.50M_0$, Cl:DOC=2:1.

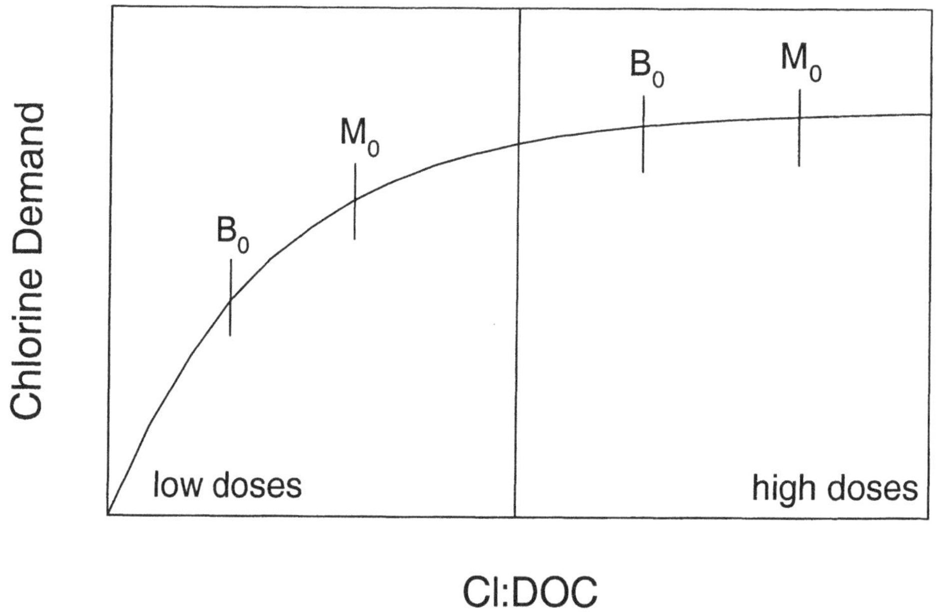

Figure 4.25: Representation of chlorine demand as a function of initial chlorine dose.

4.4.3 Results and Discussion: Phase II.b

Figures 4.26 through 4.30 show the demand study data for the South Norwalk, Cincinnati, Midland, and Las Vegas (pH 8.0 and 9.8) waters, respectively. To determine the lowest Cl:DOC to maintain a 0.2 mg/L the data is described by a saturation relationship between chlorine demand, C_d, and dose (represented by Cl:DOC)

$$C_d = \frac{K_1 * (\mathrm{Cl:DOC})}{K_2 + (\mathrm{Cl:DOC})} \tag{4.11}$$

where K_1 (mg/L) is the maximum demand at a particular time (24, 72, or 120 hours) and K_2 (mg/mg), in the context of Monod kinetics, is the "half-velocity constant," or the Cl:DOC at one half of the maximum demand at a particular time. The doses are chosen such that $C_0 - C_d \approx 0.2$ mg/L at 24 or 72 hours for B_0 and at 120 hours for M_0. Table 4.4 shows the values of K_1 and K_2, as well as the Cl:DOC, used to determine M_0 and B_0 for each water.

From these results, the demand can be calculated for the conventional dose M_0 at both boosting times. The demand from M_0 can be compared to B_0 at a particular boost time to determine if there will be a difference in chlorine demand between the two. If there is a difference, booster disinfection should initially provide a chemical kinetic advantage after boosting.

The results for these demand studies indicate that the waters from South Norwalk and Las Vegas (both pH 8.0 and 9.8) (Figures 4.26, 4.29, and 4.30) should produce a chemical

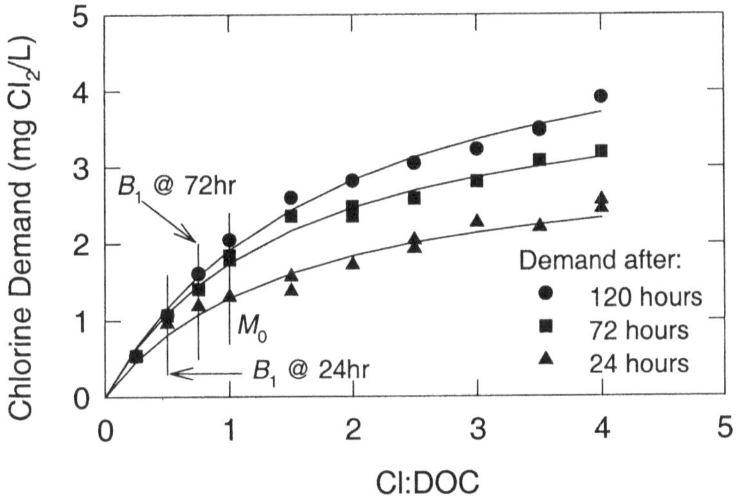

Figure 4.26: South Norwalk demand study.

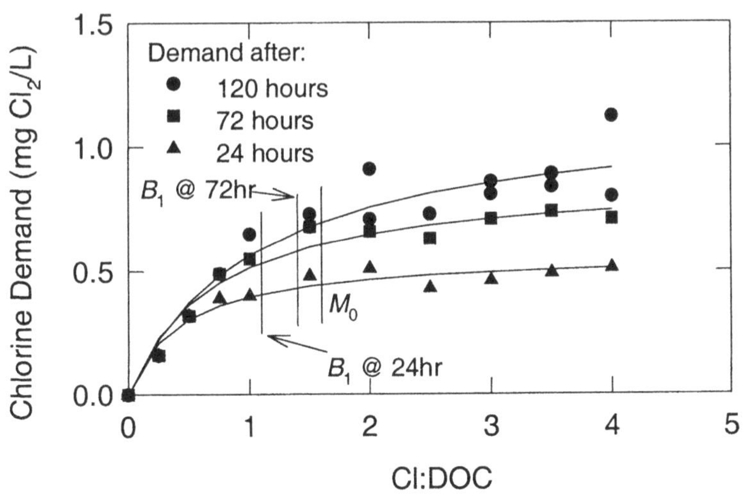

Figure 4.27: Cincinnati demand study.

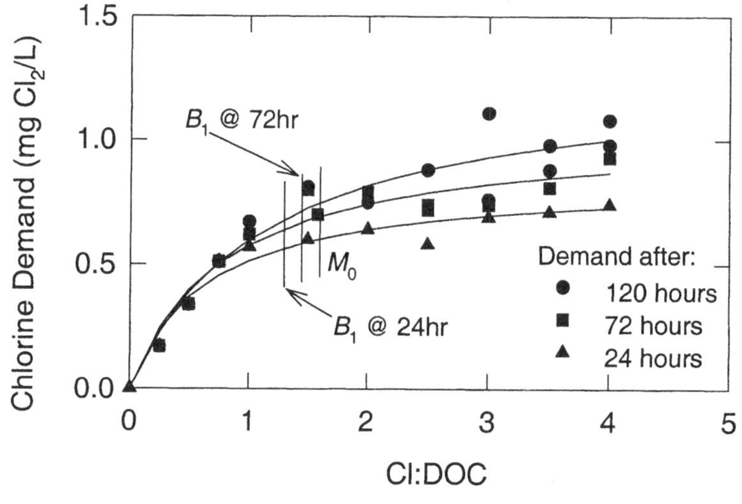

Figure 4.28: Midland demand study.

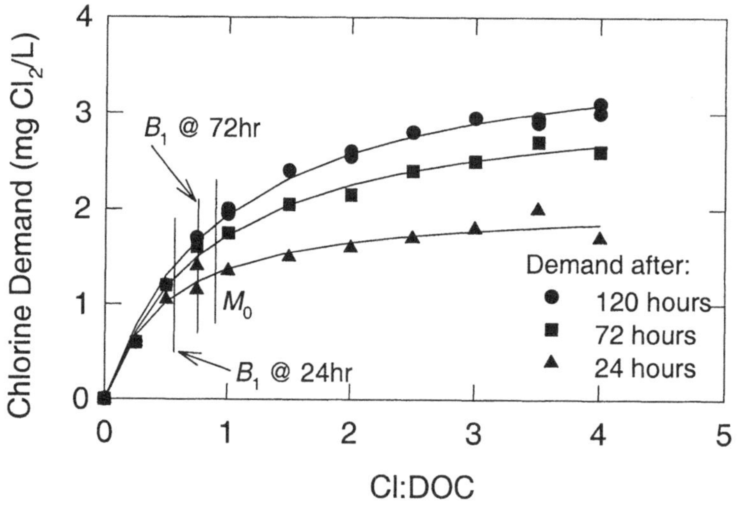

Figure 4.29: Las Vegas demand study (pH=8.0).

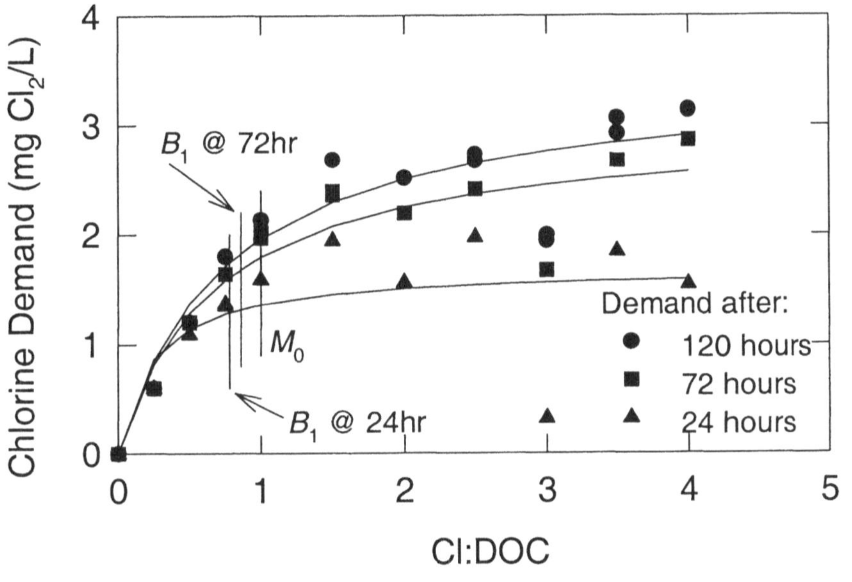

Figure 4.30: Las Vegas demand study (pH=9.8).

Table 4.4: Saturation model parameters for the different waters at various reaction times and chlorine doses.

Source	Time	Cl:DOC	K_1	K_2	R^2
South Norwalk	24	0.50:1	3.18	1.46	0.96
	72	0.75:1	4.24	1.43	0.99
	120	1:1	5.40	1.81	0.99
Cincinnati	24	1.1:1	0.56	0.42	0.95
	72	1.4:1	0.88	0.70	0.97
	120	1.6:1	1.15	1.03	0.91
Midland	24	1.3:1	0.85	0.64	0.95
	72	1.4:1	1.04	0.81	0.95
	120	1.6:1	1.29	1.16	0.93
Las Vegas (pH=8.0)	24	0.56:1	2.08	0.51	0.97
	72	0.75:1	3.24	0.87	0.99
	120	0.90:1	3.83	0.98	0.99
Las Vegas (pH=9.8)	24	0.78:1	1.68	0.24	0.41
	72	0.86:1	2.99	0.67	0.85
	120	1:1	3.46	0.77	0.88

advantage after booster disinfection, since the Cl:DOC doses used are relatively low, thus the disinfectant decay is dependent on concentration. The Cincinnati and Midland demand studies (Figures 4.27 and 4.28) indicate the demand will be independent of disinfectant dose, thus no advantage should be observed.

South Norwalk

The experimental data corresponding to single boost scenarios will be discussed in this section. The experimental data for the multiple boost scenario is presented in Appendix C. The corresponding discussion, however, is relevant to all boost dose scenarios.

Figures 4.31 and 4.32 illustrate a chemical advantage for booster disinfection in reducing the amount of disinfectant applied to maintain an adequate residual with respect to conventional disinfection. This advantage agrees with the results observed from the demand study. The associated THM formation, however, shows little, if any, *long-term* reduction in THM concentration. Reducing the initial dose 25% (Figure 4.31) did not produce less THMs than the conventional dose M_0 at any time. When the initial boost doses were 50% of M_0, however, there was an initial decrease in THM formation until the remaining mass was added. After the mass equivalent booster dose was applied, the THM formation proceeded to increase until similar THM concentrations were observed between the conventional and booster scenarios. This indicates the reaction pathways which produce a residual maintenance advantage may not be those responsible for THM formation, i.e., oxidation for the former and substitution for the latter. Figure 4.33 also confirms the existence of a chemical advantage by illustrating that the chlorine demand between the conventional and initial boost doses are dependent on concentration.

Cincinnati

The experimental data corresponding to single boost scenarios, with B_0=0.87M_0, will be discussed in this section. The experimental data for the single (B_0=0.68M_0) and multiple boost scenario is presented in Appendix C The corresponding discussion, however, is relevant to all boost dose scenarios.

Figure 4.34 illustrates booster disinfection cannot provide a chemical advantage for reducing the amount of disinfectant applied to maintain an adequate residual with respect to conventional disinfection for this water. This lack of advantage corresponds with the results from the demand study. There was no THM reduction observed, regardless of the difference in initial concentration. There also appears to be no increase in THM formation when the boost doses are applied, indicating that THM formation, for this water, may also be independent of chlorine concentration. Figure 4.35 confirms the absence of a chemical advantage by showing no difference between the chlorine demand of the conventional and initial boost doses.

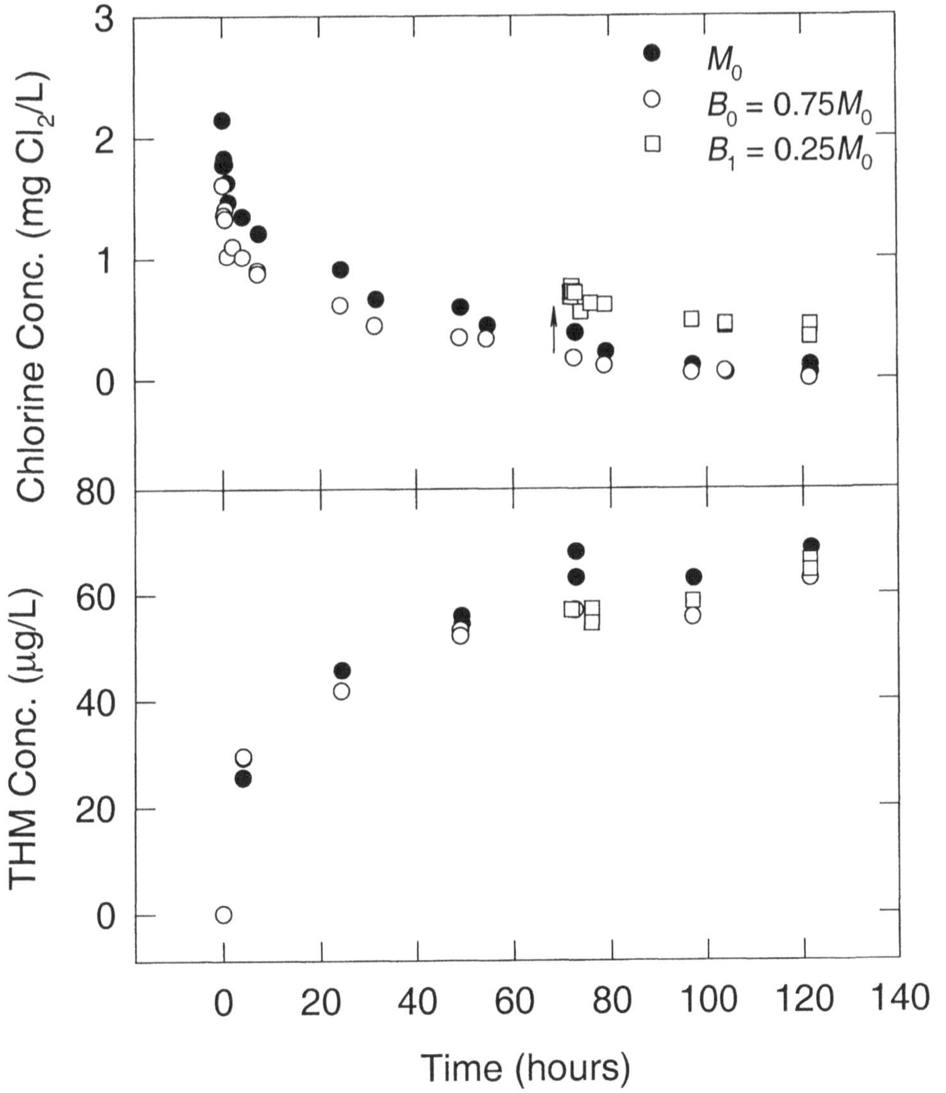

Figure 4.31: Chlorine decay and THM formation data of phase IIb for South Norwalk water; $B_0 = 0.75M_0$.

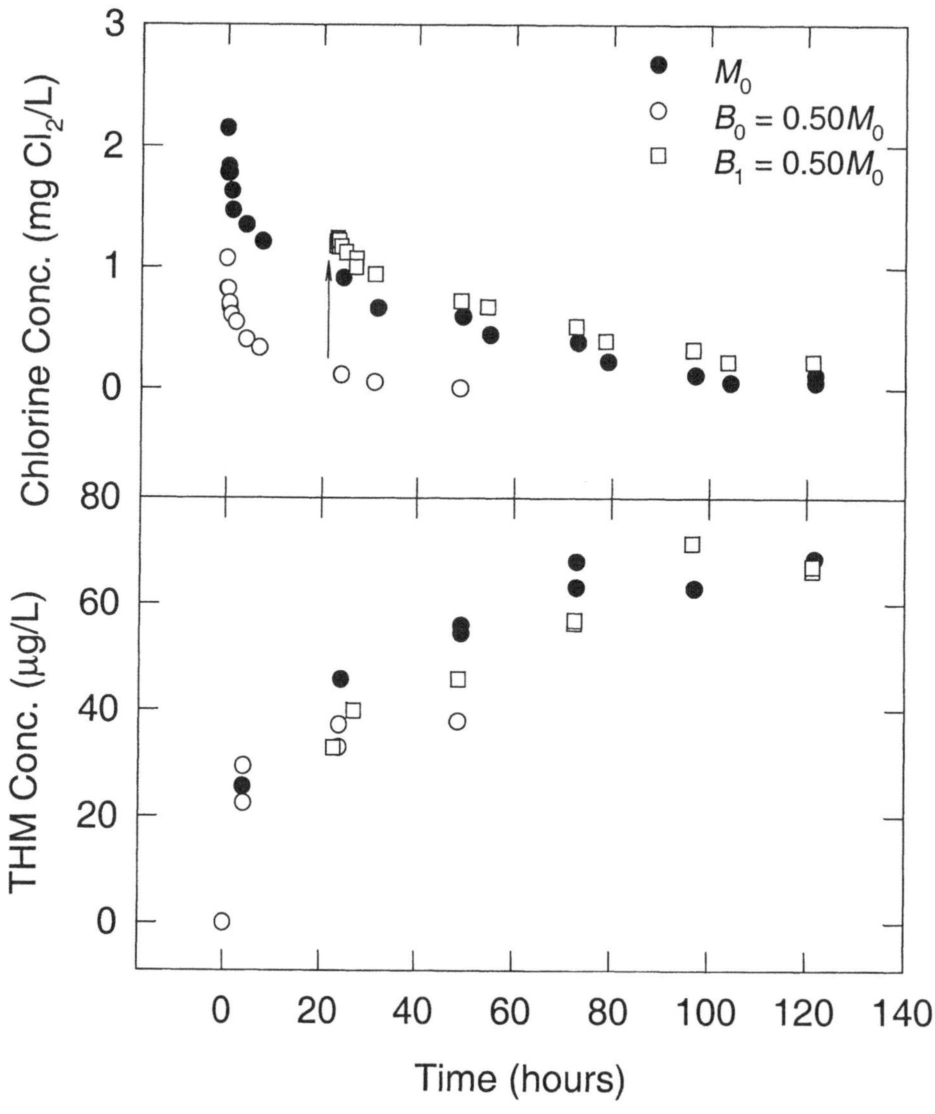

Figure 4.32: Chlorine decay and THM formation data of phase IIb for South Norwalk water; $B_0 = 0.5M_0$.

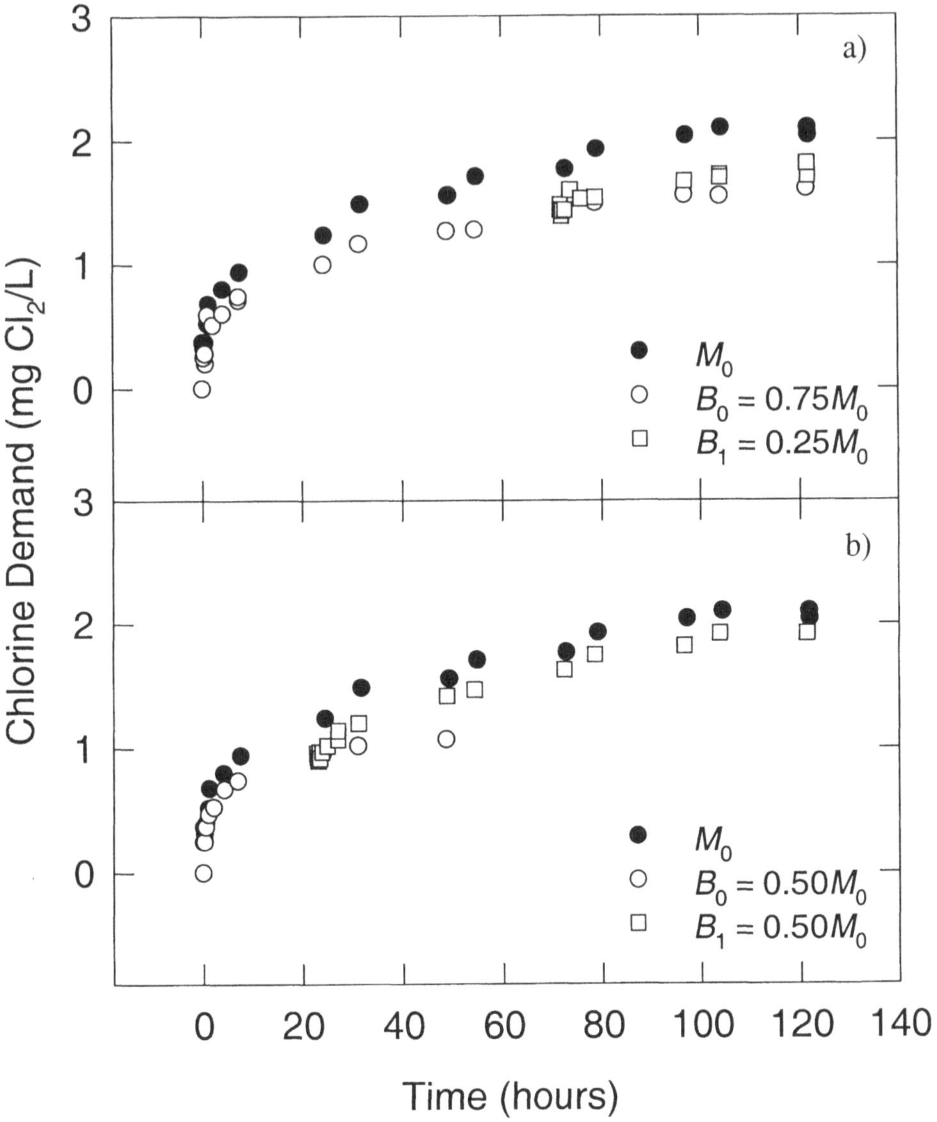

Figure 4.33: Chlorine demand data of phase IIb for South Norwalk water; a) $B_0 = 0.75M_0$, b) $B_0 = 0.5M_0$.

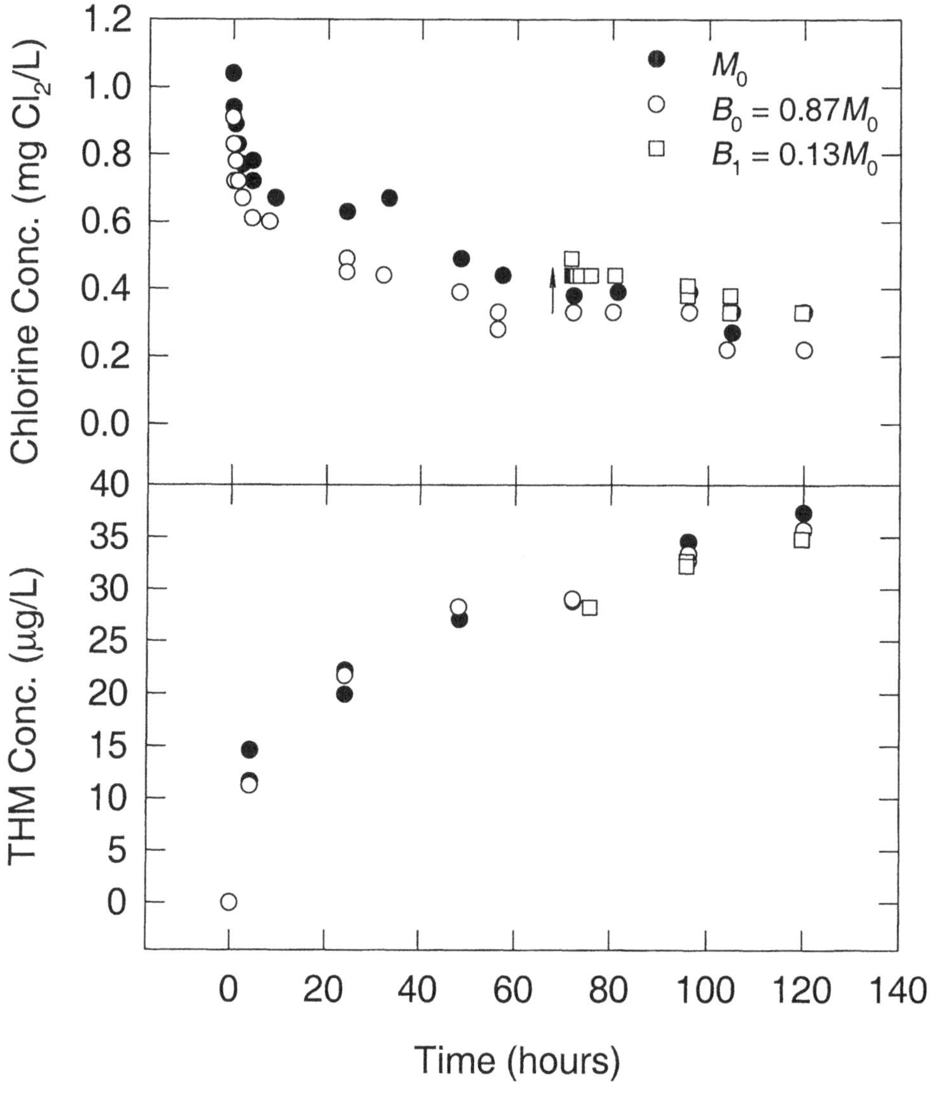

Figure 4.34: Chlorine decay and THM formation data of phase IIb for Cincinnati water; $B_0 = 0.87M_0$.

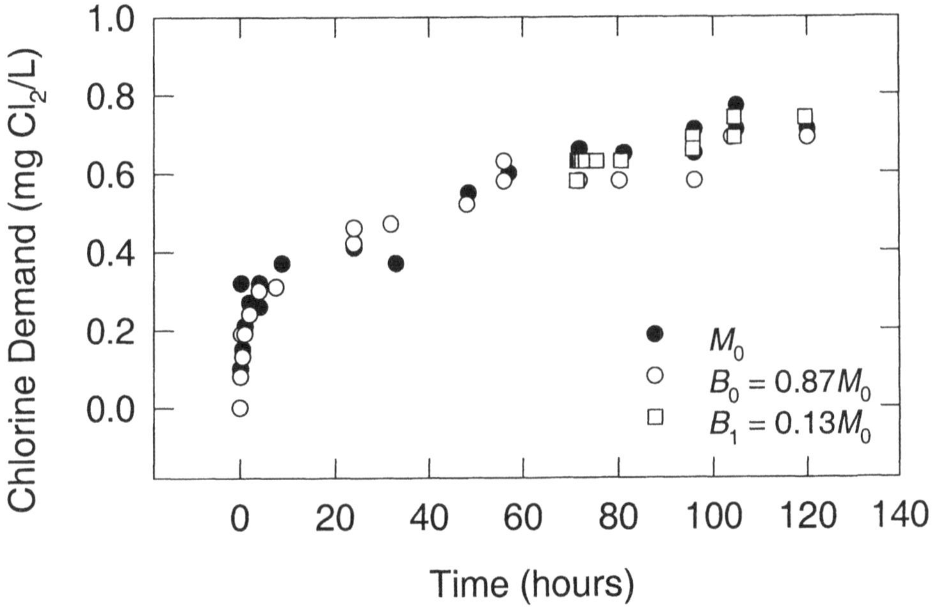

Figure 4.35: Chlorine demand data of phase IIb for Cincinnati water; $B_0 = 0.87M_0$.

Midland

Figures 4.36 and 4.37 illustrate booster disinfection cannot provide a chemical advantage for reducing the amount of disinfectant applied to maintain an adequate residual with respect to conventional disinfection in this water. The observed experimental decay agreed with the predictions from the demand study. The associated THM formation, however, appears to exhibit a slight long-term reduction in THM formation. Reducing the initial dose 13% (Figure 4.36) did not produce any less THMs than the conventional dose M_0, however, after applying the booster dose there appeared to be less THM formation than for conventional chlorination. When the initial boost doses were 18% of M_0 there was a decrease in THM formation until the time of mass equivalency, after which the THM formation again appeared to increase. As with the South Norwalk water, this may indicate the reaction pathways associated with disinfectant decay may not be the same pathways responsible for THM formation. Additionally, groundwater sources, like Midland, typically have relatively high inorganic concentrations (including Br^-), which may exhibit higher chlorine demands than surface waters through reaction with the inorganic material. These reactions with inorganic compounds do not form THMs (with the significant exception of Br^-), and may be responsible for the reduction in THM formation under booster disinfection, even though there is similar chlorine demand with respect to conventional disinfection. Figure 4.38 confirms the absence of a chemical advantage by showing the chlorine demand of the conventional and initial boost doses are independent of concentration.

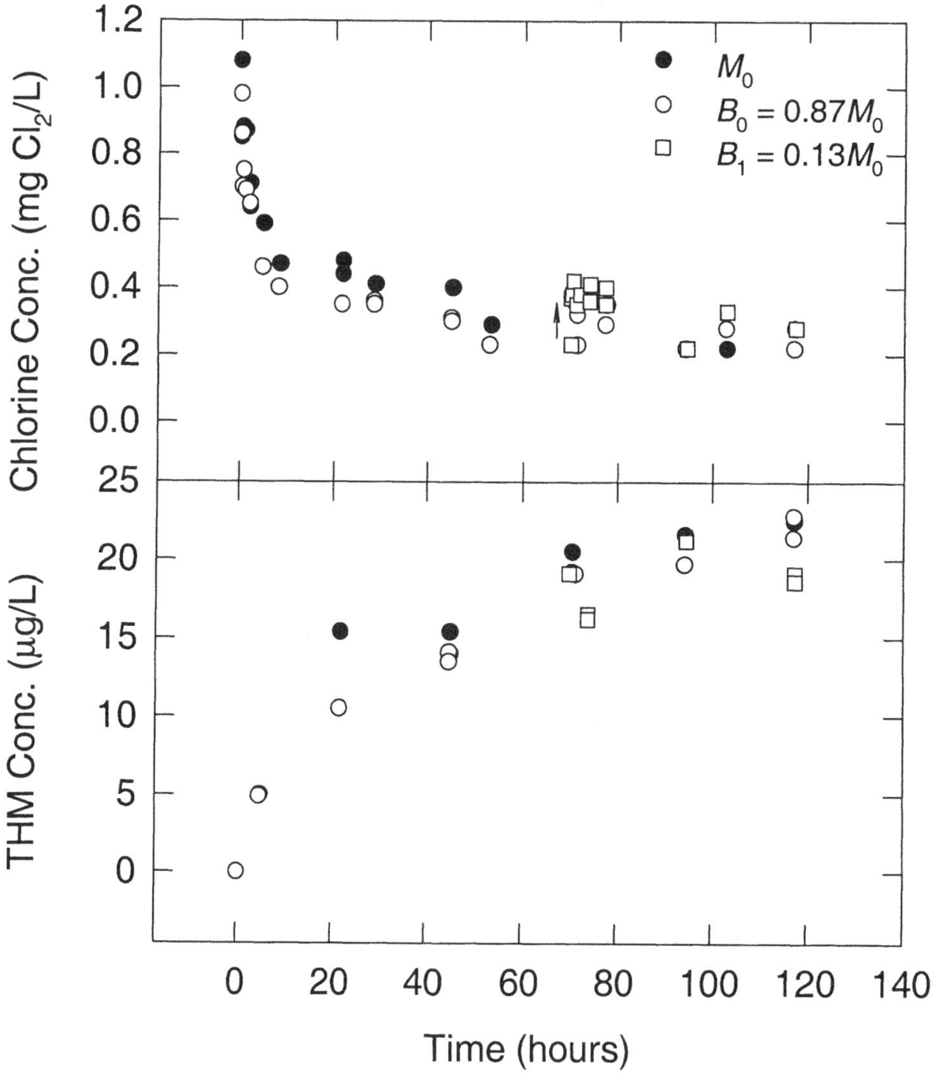

Figure 4.36: Chlorine decay and THM formation data of phase IIb for Midland water; $B_0 = 0.87M_0$.

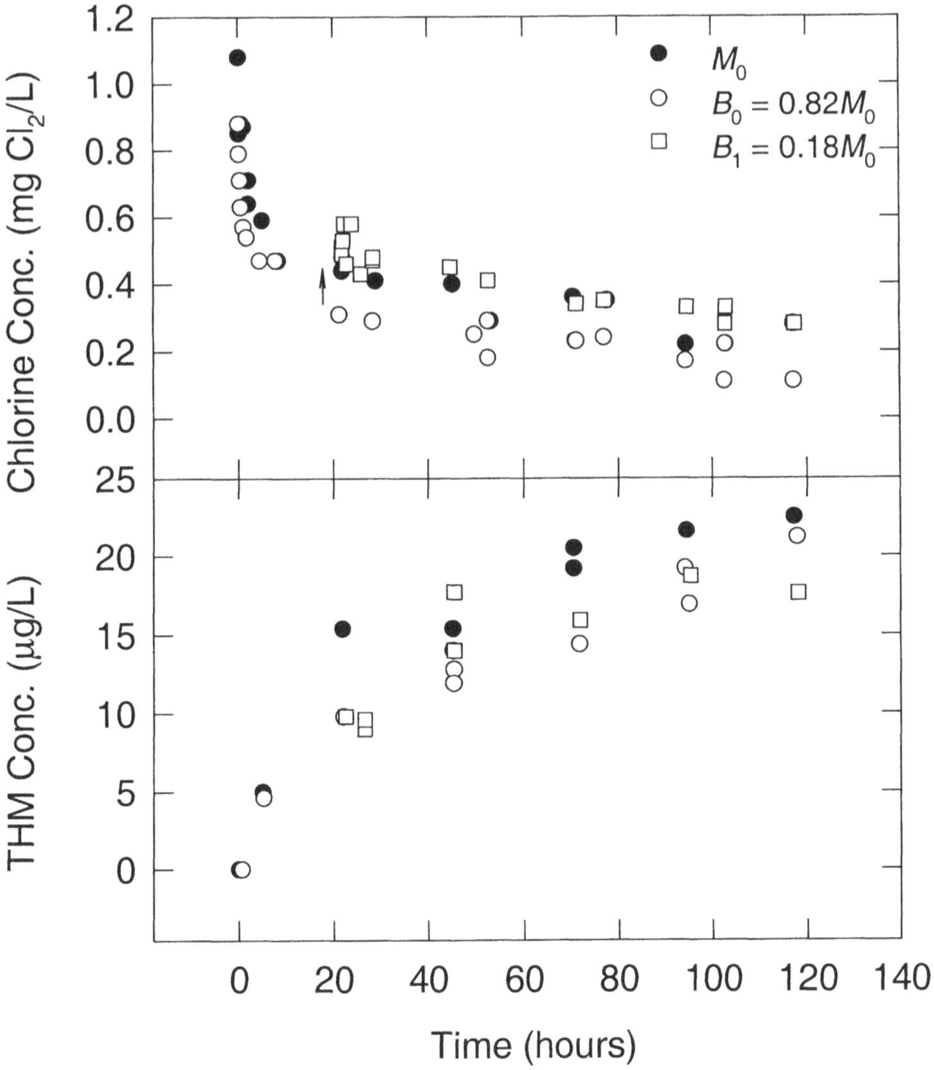

Figure 4.37: Chlorine decay and THM formation data of phase IIb for Midland water; $B_0 = 0.82M_0$.

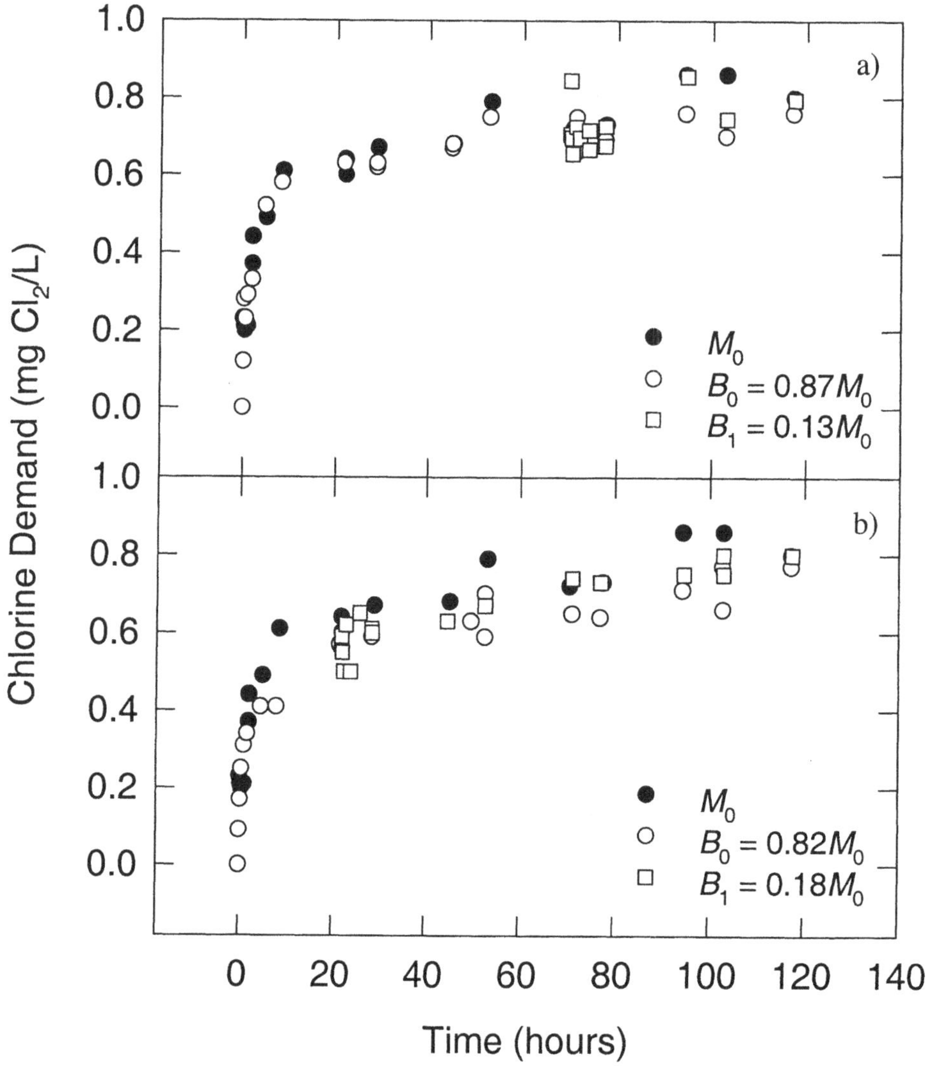

Figure 4.38: Chlorine demand data of phase IIb for Midland water; a) $B_0 = 0.87M_0$, b) $B_0 = 0.82M_0$.

Table 4.5: Final THM concentrations for various disinfection scenarios and Las Vegas water, pH=8.0 and pH=9.8.

pH	M_0	$B_0 + B_1$, 24 hrs.	$B_0 + B_1$, 72 hrs.
8.0	82.6	73.7	77.8
9.8	103	96.8	97.4

Las Vegas - pH 8.0 and 9.8

Figure 4.39 shows booster disinfection can provide a chemical advantage immediately after boosting at pH 8.0, however, this advantage disappears as the reaction continues (these results are similar to those for $B_0 = 0.83M_0$ in Appendix C). This initial advantage corresponds with the results from the demand study, and reinforces that the disinfectant decay after the boost cannot be estimated by using the demand study. Figure 4.40 shows no kinetic advantage at pH 9.8 even though the demand study predicted a slight initial advantage for $B_0 = 0.86M_0$. There was no reduction in THM formation for any of the booster disinfection scenarios, however, there was a difference between the two pH values. Table 4.5 illustrates the difference in THM formation between disinfection scenarios at a single pH is insignificant, however, increasing the pH increases the THM yield (THM formation is base promoted, see Figures 4.2 and 4.3). These results also indicate that the reaction pathways associated with disinfectant decay may not exactly correspond to those responsible for THM formation. Figure 4.41 confirms the existence of a slight chemical advantage for $B_0 = 0.62M_0$, showing that the chlorine demand between the conventional and initial boost doses are dependent on concentration. Figure 4.42 illustrates the disinfectant decay at pH 9.8 is independent of disinfectant concentration.

4.4.4 Summary

Under an equivalent mass scenario, the potential advantages of booster disinfection are dependent on the initial disinfectant doses. Under relatively high disinfectant concentrations, the decay of the conventional and initial boost dose are relatively independent of disinfectant concentration; thus, when the boost scenario is made mass equivalent, the resulting disinfectant concentration is the same as the conventional scenario and occurs at a similar chlorine demand (indicating the reactions of both scenarios have proceeded to a similar degree). For relatively low initial disinfectant doses, the decay of the conventional and initial boost doses appear dependent on disinfectant concentration; thus, the initial boost dose decays less than the conventional dose resulting in a higher concentration when mass equivalency is achieved. The increase in concentration immediately after the boost dose does not, however, guarantee a long term benefit as the resulting chlorine decay is a poorly understood function of dis-

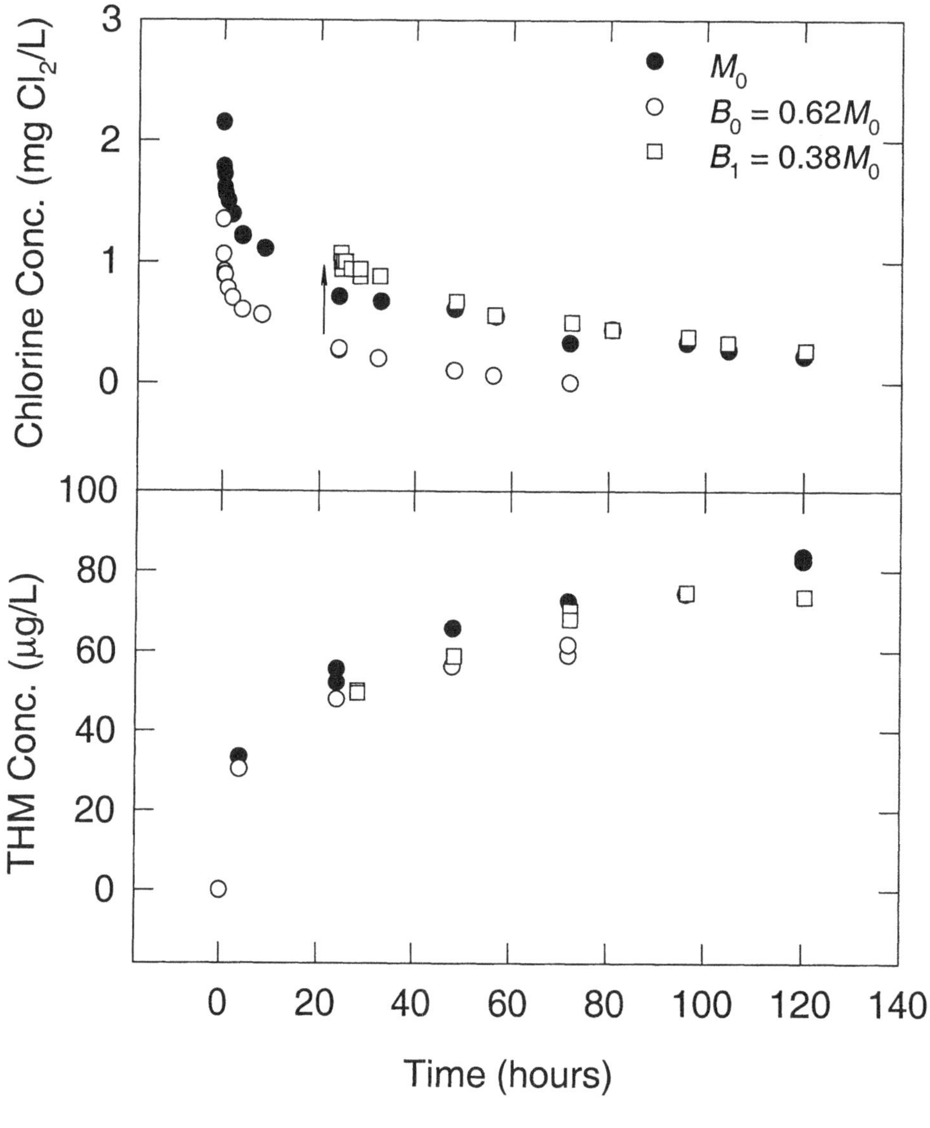

Figure 4.39: Chlorine decay and THM formation data of phase IIb for Las Vegas water (pH=8.0); $B_0 = 0.62M_0$.

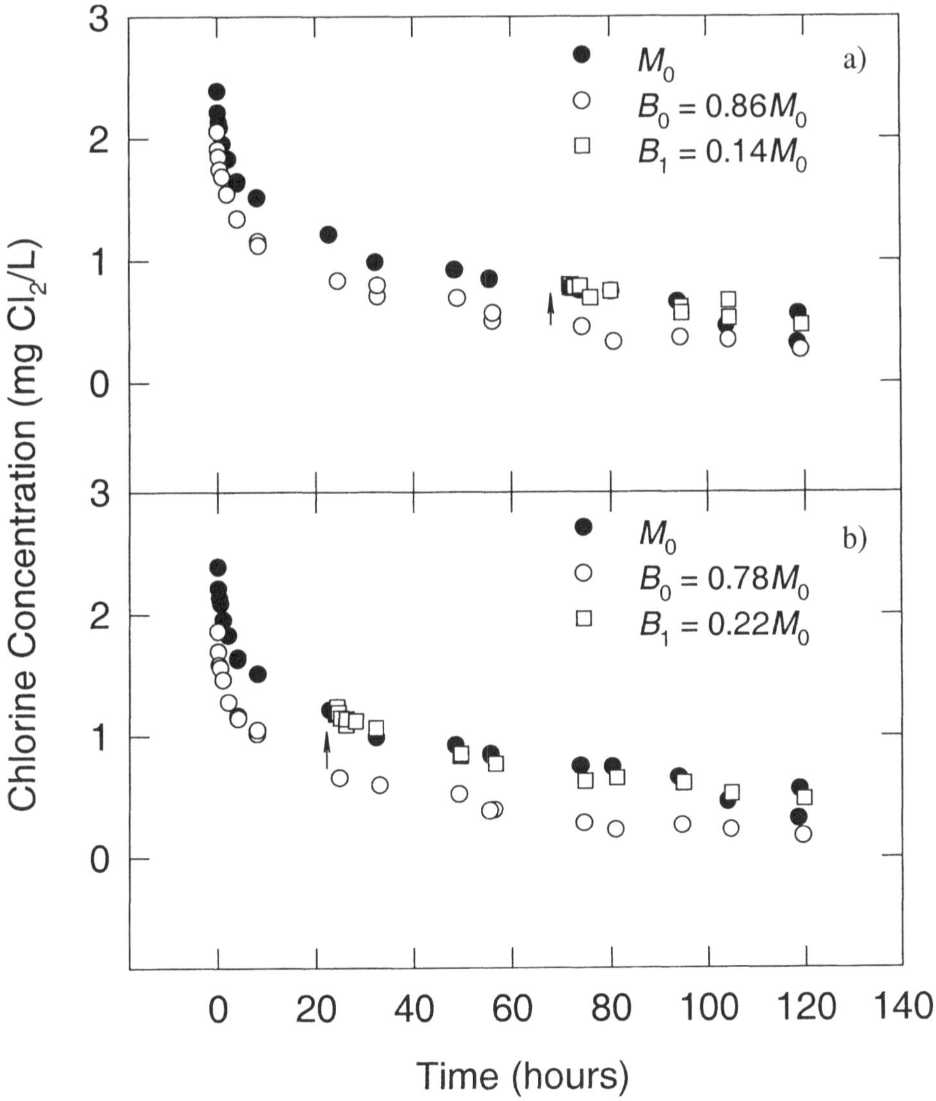

Figure 4.40: Chlorine decay data of phase IIb for Las Vegas water (pH=9.8); a) $B_0 = 0.86M_0$, b) $B_0 = 0.78M_0$.

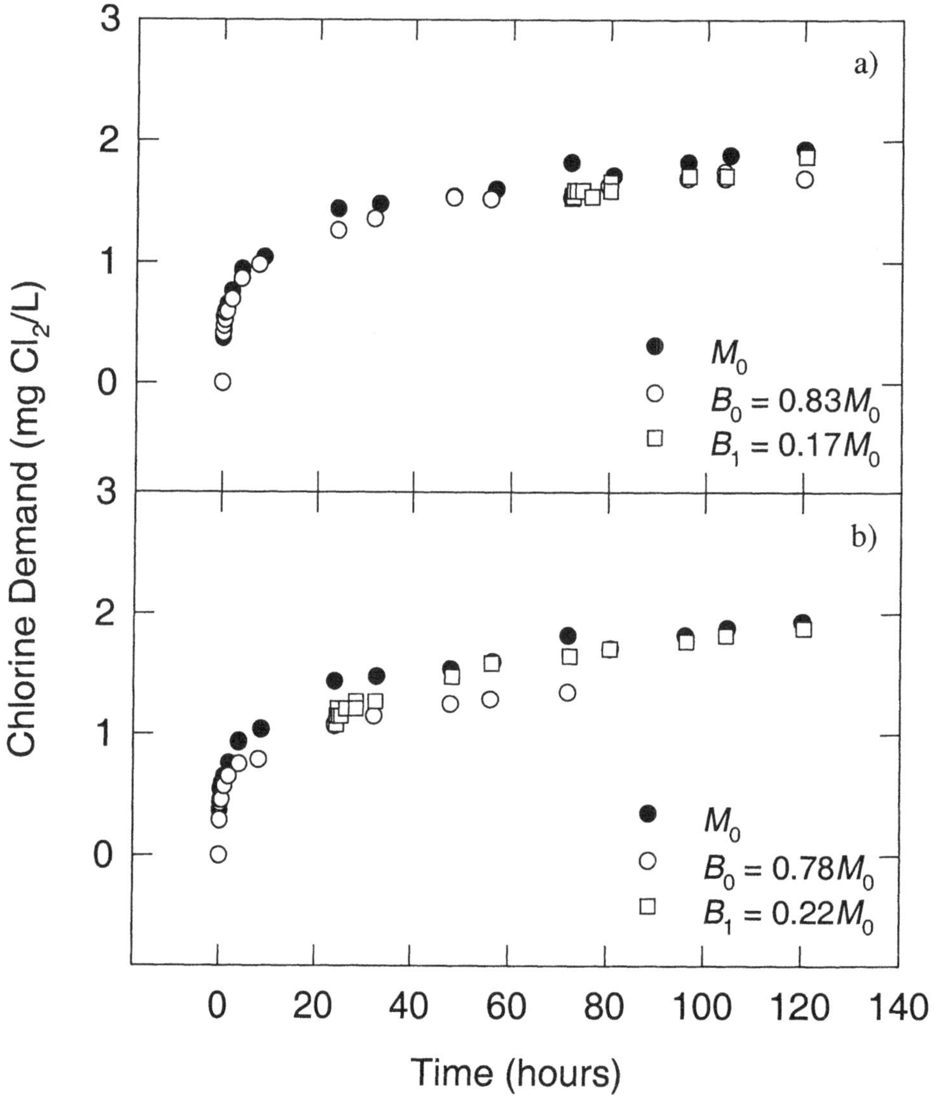

Figure 4.41: Chlorine demand data of phase IIb for Las Vegas water (pH=8.0); a) $B_0 = 0.83M_0$, b) $B_0 = 0.62M_0$.

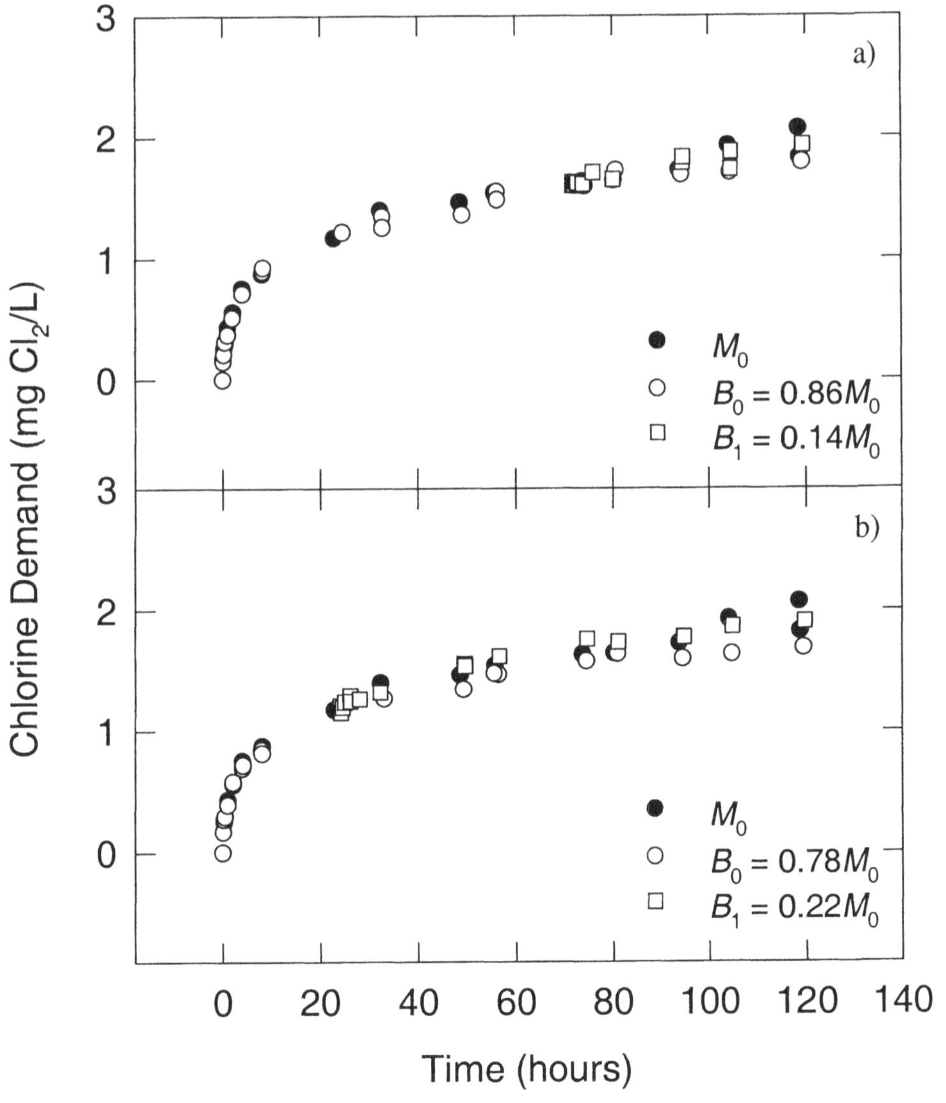

Figure 4.42: Chlorine demand data of phase IIb for Las Vegas water (pH=9.8); a) $B_0 = 0.86M_0$, b) $B_0 = 0.78M_0$.

infectant concentration and reactive material. The notion of what constitutes a "relatively high" or "relatively low" dose is water–specific.

For the four waters examined, using equivalent mass booster disinfection to reduce the average disinfectant concentration did not reduce the THM formation. This suggests THM formation is independent of disinfectant concentration under the range of experimental conditions studied. Reduced THM formation was observed, however, for some waters prior to the systems being mass equivalent, although the reduced THM concentration disappeared after the systems were made mass equivalent. While the final THM concentration was not reduced, the initial reduction in THM concentration would reduce the THM exposure of consumers receiving pre-boost water. For those waters in which booster disinfection could reduce the total amount of disinfectant applied to the system, performing experiments which actually reduce the applied mass (instead of mass equivalency) may show a reduction in THM formation under booster chlorination.

4.5 Estimating Disinfectant Mass Savings

This section provides a detailed methodology for determining the potential of a given water to consume less disinfectant mass under booster chlorination, while maintaining an adequate residual. This method only determines if an advantage may be achieved immediately after the boost dose; the magnitude of any long-term benefits must be determined by disinfectant decay studies. The following procedure will, however, determine the initial dose concentrations necessary to maintain a specified residual for use in performing the more detailed disinfectant decay experiments.

4.5.1 Preparation

The following is list of necessary equipment:

1. glassware - amber glass bottles with a minimum volume of 125mL
2. DOC measurement device
3. pH meter
4. disinfectant dosing solution – add pH 6.7 borate buffer to chlorine solution (1000-1500 mg Cl_2/L to yield a pH 8.0 dosing solution (approximately a 4-5:1 volume ratio of pH 11.2 hypochlorite solution to pH 6.7 borate buffer; must account for a 20-25% reduction in hypochlorite strength due to dilution)(Summers et al., 1996)
5. pH 6.7 borate buffer solution - 1.0 M (61.83 g/L) boric acid (ACS grade) and 0.11 M (4.4 g/L) sodium hydroxide (ACS grade) in boiled laboratory clean water (RO/IX/GAC)(Summers et al., 1996)

6. pH 8.0 borate buffer solution - 1.0 M (61.83 g/L) boric acid (ACS grade) and 0.255 M (10.2 g/L) sodium hydroxide (ACS grade) in boiled laboratory clean water (RO/IX/GAC)(Summers et al., 1996)
7. chlorine measurement equipment, dosing solution - Method 4500-Cl B(APHA et al., 1995)
8. chlorine measurement equipment, residual concentration - DPD-FAS Method 4500-Cl F(APHA et al., 1995)
9. constant 20°C storage facility
10. source water – preferably full scale treated water before disinfectant addition (includes pre-treatment); if not possible, the water should be pilot/bench scale treated to mimic full scale treatment.

Before starting the experiments three key experimental variables must be determined: 1) the desired minimum residual, 2) the number and time of the boost doses, and 3) the ultimate length of the experiment. Determining the appropriate boost times may require a significant amount of consideration. The boost dose times should be representative of the residence time of the water required to reach the (potential) booster station; thus, estimating the residence time prior to locating the booster stations is a challenging problem that will likely require an iterative approach. The ultimate time is meant to represent the longest residence time in the distribution system.

Prior to the experiment, all glassware should be cleaned to be chlorine demand free following the UFC conditions specified in Summers et al.(Summers et al., 1996). If the collected treated source water is not used immediately, the water should be stored at a constant 4°C in the dark to minimize water quality degradation. The DOC of the water should be measured on two samples, in duplicate, following the procedures of the DOC measurement instrument. Finally, the water should then be buffered to pH 8.0, using the pH 8.0 borate buffer, and the chlorine dosing solution strength measured.

4.5.2 Experimental

The initial disinfectant dose to maintain an adequate residual at the various boost times, and ultimate time, is desired. To determine the appropriate dose, several samples of the source water are separated into the chlorine demand free amber bottles and dosed to different Cl:DOC for each of the predetermined times. For each set of Cl:DOC based on time, 30% of the samples are chosen to be run in duplicate. The water sources are then dosed with the appropriate amount of chlorine to achieve the desired Cl:DOC and allowed to react in the dark at a constant 20°C until the predetermined time is reached. At each of the predetermined times, the disinfectant residuals of the various Cl:DOC are measured and the demand determined (demand = initial concentration - residual concentration).

While the exact number of Cl:DOC examined per time will vary from system to system, the Cl:DOC of the South Norwalk Phase II.b experiment will be used as an example. For this example, the boost times were chosen as 24 and 72 hours, with an ultimate time of 120 hours. A set of ten Cl:DOC were chosen 0.25, 0.50, 0.75, 1.0, 1.5, 2.0, 2.5, 3.0, 3.5, and 4.0:1, with three duplicates chosen from the initial set of ten Cl:DOC. For each boost time (24, 72, and 120 hours), a set of the ten Cl:DOC and three duplicates were prepared. As each sample time was reached, the entire set of Cl:DOC samples and duplicates were measured for pH and chlorine residual. The resulting data is shown in Figure 4.26; notice each curve represents the chlorine demand at one of the given times for all of the initial concentrations based on the various Cl:DOC.

While these experimental conditions worked well for the four waters investigated (DOC from 0.65 to 2.15 mg/L and an ultimate time of 120 hours), the experimental conditions may vary from source to source with varying boost times, ultimate time, and decay characteristics of the water. For water with large decay rate, larger Cl:DOC may be required. Additionally, increasing the boost times, or ultimate time, may also require larger Cl:DOC to ensure maintenance of an adequate residual. To accomodate larger Cl:DOC doses, additional Cl:DOC may be used initially, or the differences between the ten individual Cl:DOC may be increased to keep the number of samples the same. Similar reductions may occur should the boost, or ultimate, times be shorter than in the example.

4.5.3 Analysis

Once the experimental data is gathered, the initial doses required to achieve the desired minimum residual may be estimated using (4.11). For each set of demand data collected at each particular sample time, the parameters of (4.11) may be estimated by minimizing $\sum_{i=1}^{n}(C_i^m - C_i)^2$ using a non-linear solver (e.g. EXCEL's Solver function), where n is the number of Cl:DOC used in the demand study, and C_i^m is the modeled chlorine demand (from (4.11) and C_i is the experimental chlorine demand of the dose based on the ith Cl:DOC ratio. At each particular time, the Cl:DOC which provides an adequate residual must be determined such that $C_0 - C_d =$ desired residual, where C_0 is the initial chlorine concentration and C_d is the chlorine demand at a particular time for a given Cl:DOC.

Given the optimal parameter values K_1 and K_2, (4.11) may be rewritten

$$C_0 - \frac{K_1 * (\mathrm{Cl:DOC})}{K_2 + (\mathrm{Cl:DOC})} = \text{residual} \tag{4.12}$$

The Cl:DOC which satisfies this relationship, $\mathrm{Cl:DOC_{opt}}$, is determined

$$\mathrm{Cl:DOC_{opt}} = \frac{K_2 \times (\mathrm{Cl:DOC} - \text{residual})}{K_1 - (\mathrm{Cl:DOC} - \text{residual})} \tag{4.13}$$

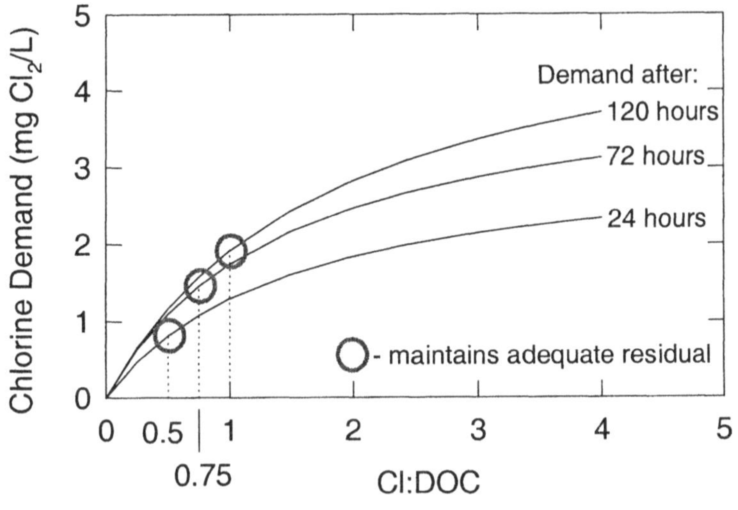

Figure 4.43: Modeled chlorine demand and $Cl : DOC_{opt}$ values of the South Norwalk water.

The chlorine demand for $Cl : DOC_{opt}$ of a particular boost time is compared to that for $Cl : DOC_{opt}$ of the ultimate time, with the difference being the estimate of the potential initial benefit. A difference of 0.1 mg/L, or less, will be difficult to detect with any certainty.

For the example South Norwalk data, Figure 4.26 shows the modeled chlorine demand from (4.11) as well as the observed experimental data. Figure 4.43 illustrates the resulting $Cl : DOC_{opt}$ for the boost times of 24 and 72 hours, and the ultimate time of 120 hours (without the observed experimental data for clarity; Figure 4.26 contains similar data with the observed experimental data). The potential benefit of booster disinfection immediately after the boost is determined by the difference in chlorine demand of the $Cl : DOC_{opt}$ for the ultimate time, after 24 hours of decay, and the $Cl : DOC_{opt}$ for the 24 hour boost time and is illustrated in Figure 4.44 to be approximately 0.5 mg/L. A similar analysis can be carried out at the boost time of 72 hours to show an initial benefit of 0.3 mg/L. Since the potential benefits immediately after the boost appear to be significant, a more detailed, long-term decay experiment would be performed using initial chlorine concentrations which correspond to the determined $Cl : DOC_{opt}$. If there was no apparent immediate advantage for booster disinfection, the more detailed experimental decay study would not be required.

While this procedure may increase the time to perform an experimental decay study, there are two distinct advantages. The first advantage of this method is the ability to determine the initial doses which meet some residual concentration at a specific time. Such accuracy would be difficult to determine without the prior knowledge afforded by the demand study. Secondly, several boost times may be evaluated, relatively quickly when compared to a detailed experimental decay study, allowing a determination to be made about the necessity of running a dctailed disinfectant decay study.

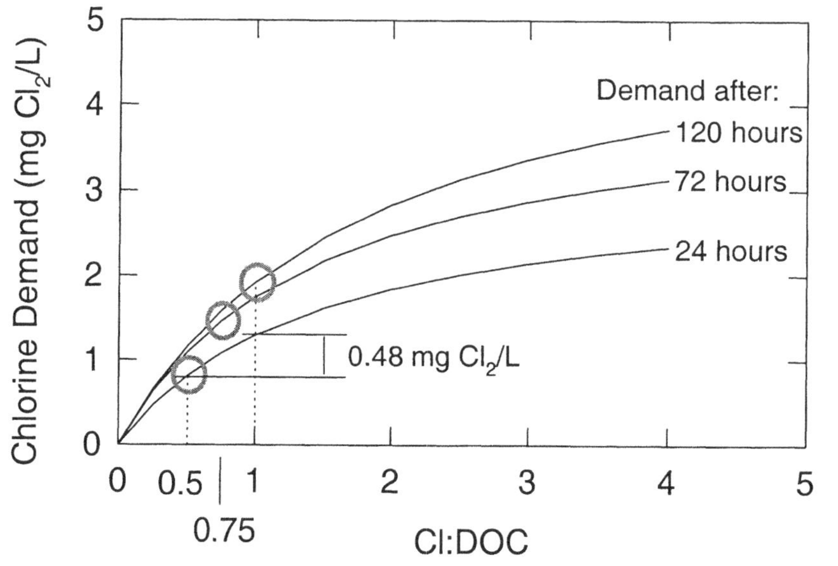

Figure 4.44: Modeled chlorine demand and $\mathrm{Cl : DOC_{opt}}$ values of the South Norwalk water.

4.5.4 Summary

A procedure was developed to estimate the potential of booster disinfection to reduce the amount of disinfectant applied to the system. In this procedure, disinfectant decay for several different Cl:DOC ratios are observed to determine the dose required to maintain a residual at two different times: 1) the maximum residence time of the distribution system, and 2) the residence time until the (first) boost dose. If the demand of the two disinfectant doses at the time of boosting are equivalent, no benefit is expected. If a difference in demand does exist, however, a benefit immediately after boosting is expected. The increased potential for waters with higher DOC is due to the larger range of disinfectant doses in which the decay is dependent on concentration. This procedure alone can not determine if the reduced demand will persist, or if reduced THM formation may occur as a side benefit.

4.6 Alternative Disinfectants: Chloramines

A preliminary chloramine study was performed using treated Portland water to examine chloramine decay under booster chlorination. The focus of the experiment was to determine if booster chlorination could: 1) reduce the amount of ammonia formed in the distribution system (see (Valentine et al., 1997)), and 2) reduce the average ammonia concentration over the experiment duration. This emphasis on ammonia formation is in response to utility concerns regarding nitrification. The amount of ammonia present may be used as a surrogate for the likelihood of nitrification.

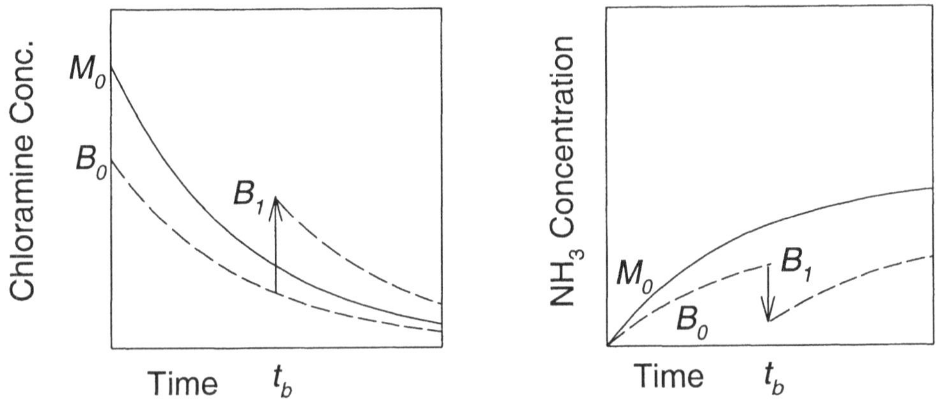

Figure 4.45: Illustration of the experimental design for a single booster chlorination scenario with B_1 based on a Cl:N of 5:1 or 3:1 at the time of boosting.

4.6.1 Experimental Design

The DOC of the treated Portland water was 1.4 mg/L. M_0 and B_0 were based on Cl:DOC determined by demand studies over 14 days (the length of the experiment) and 7 days (the time of boosting doses), respectively, to maintain a residual of 2.0 mg/L. The longer experimental durations are in consideration of nitrification episodes occurring in locations with relatively long residence times(Wolfe et al., 1988) and nitrifying bacteria being slow-growers(Geldreich, 1996), thus long term ammonia concentrations may be more important than short term ammonia concentrations. The choice of a 2.0 mg/L residual was based on the median residual concentration determined from a survey of 438 utilities that maintain chloramine residuals within their distribution system(Mann et al., 1998). A full set of experiments were run with two different initial Cl:N of 3:1 and 5:1 at pH 8.0. For each experiment, a boost dose B_1 applied enough chlorine to recombine with any ammonia present in the water to a Cl:N of either 3:1 or 5:1 at $t = 7$ days (or two boost scenarios for each of the initial Cl:N). It is important to realize these experiments may not be mass equivalent as the chlorine added at the boost is dependent on the ammonia concentration at the time of boost. THM formation is typically not considered to be a concern with chloraminated systems, therefore, THM concentrations were not measured. Figure 4.45 illustrates the different single dose booster chlorination scenarios with an initial dose of chloramines.

4.6.2 Results and Discussion

Figures 4.46, and 4.47 show chloramine demand and ammonia concentration as a function of Cl:DOC for the demand studies after 7 and 14 days with an initial Cl:N of 3:1 and 5:1, respectively. Both the chloramine demand and ammonia concentration can be described by a linear equation fit through the origin, $C_d = a(\text{Cl} : \text{DOC})$ and $\text{NH}_3 = b(\text{Cl} : \text{DOC})$ and are shown in Figures 4.46, and 4.47. The doses chosen for the experiment are such that

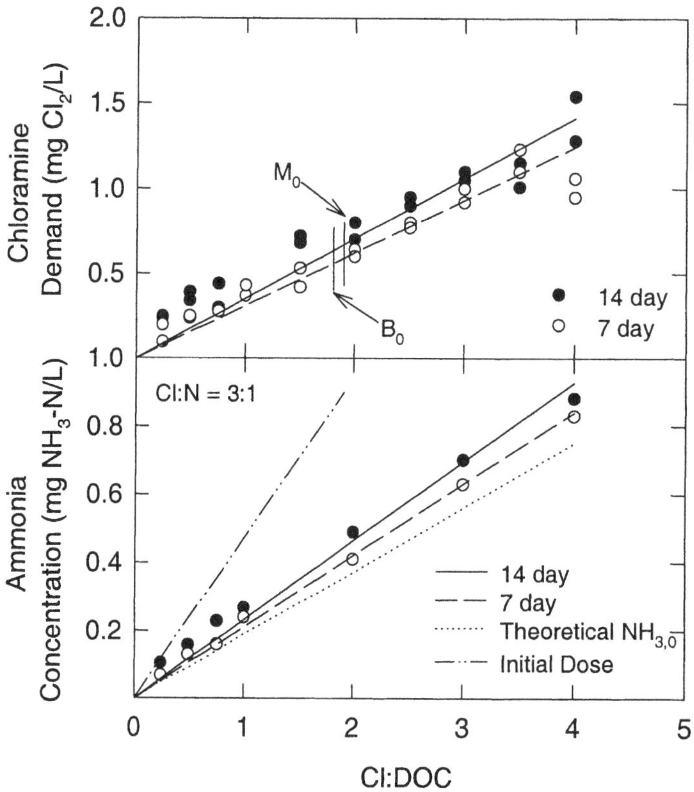

Figure 4.46: Chloramine demand and ammonia concentration from the Portland demand study for Cl:N = 3:1.

$C_0 - C_d \simeq 2.0\ mg/L$ after 7 days for B_0 and 14 days for M_0. Table 4.6 shows the values of a and b, as well as the Cl:DOC, used to determine M_0 and B_0 for the different Cl:N.

An additional line for ammonia concentration is presented in Figure 4.46 to represent the theoretical amount of ammonia not initially combined with chlorine. The theoretical initial ammonia concentration is

$$\mathrm{NH}_{3,0} = \mathrm{Cl} : \mathrm{DOC} \times \mathrm{DOC}mg/L \times \frac{1mg\mathrm{NH}_3\text{−}\mathrm{N}/L}{3mg\mathrm{Cl}_2/L} \times 0.40 \tag{4.14}$$

where 0.40 is the amount of ammonia not initially combined with chlorine ($1 - 3/5$, where 3 is the dosed Cl:N, and 5 is the stoichiometric equivalent Cl:N). No theoretical line is provided for Figure 4.47 as it should be zero. Figure 4.48 shows the change in ammonia concentration ($\Delta\mathrm{NH}_3 = \mathrm{NH}_3 - \mathrm{NH}_{3,0}$) due to chloramine decay, where NH_3 is the experimental ammonia concentration and $\mathrm{NH}_{3,0}$ is the theoretical ammonia concentration not initially combined with chlorine. The observed ammonia release due to disinfectant decay (Figure 4.48) is greater for a Cl:N of 5:1, which may be indicative of the increasing decay observed at higher Cl:N.

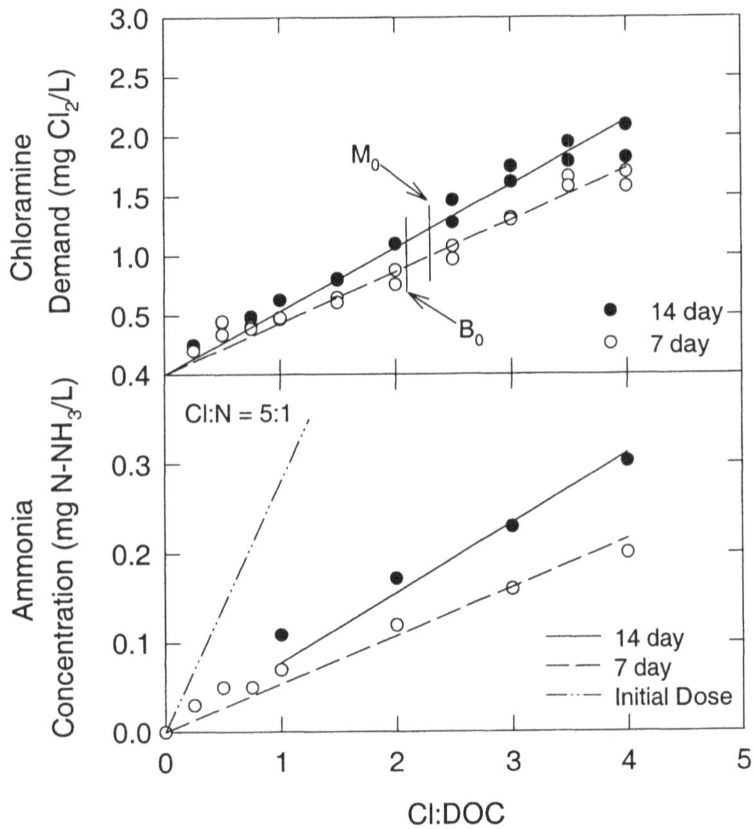

Figure 4.47: Chloramine demand and ammonia concentration from the Portland demand study for Cl:N = 5:1.

The results from the demand studies indicate chloramine decay is a function of concentration; thus, a chemical advantage may be achieved by booster disinfection when maintaining a chloramine residual. This advantage, however, will be overwhelmed when the intent of boosting with chlorine is to recombine with the ammonia concentration resulting from the initial incomplete formation of chloramine or chloramine decay and not the maintenance of a specific residual. For example, with a Cl:N of 5:1 (Figure 4.47), applying an initial boost dose (B_0) to maintain a 2.0 $mg\text{Cl}_2/L$ residual after 7 days requires 0.28 $mg\text{Cl}_2/L$ less than the associated conventional dose (M_0). A boost dose, to a Cl:N of 3:1 (which requires the least amount of chlorine), would require 0.33 $mg\text{Cl}_2/L$ to recombine the 0.11 $mg\text{NH}_3\text{-N}/\text{L}$ present after 7 days (Figure 4.46), thus nullifying any mass savings that might have been achieved. While achieving such a mass savings would be beneficial, the main focus of the experiments is to evaluate the ability of booster disinfection to reduce the amount of ammonia initially applied to the system, and to keep the average ammonia concentration at low levels. The demand study has shown the amount of ammonia initially applied can be reduced, however, it says nothing about the ammonia formation which will occur after the boost dose.

Table 4.6: Linear chlorine demand, a, and ammonia formation, b, model parameter values for the Portland water at reaction times of 7 and 14 days and Cl:N of 5:1 and 3:1.

Time (d)	Cl:N	a (mg Cl_2/L)	R^2	b (mg NH_3-N/L)	R^2
7	3:1	0.311	0.980	0.210	0.999
14	3:1	0.353	0.972	0.232	0.993
7	5:1	0.433	0.991	0.054	0.983
14	5:1	0.533	0.991	0.078	0.993

Table 4.7: Average ammonia concentration observed under the various disinfection scenarios for the Portland water.

	Ammonia ($mg$$NH_3$-N/L)		
Cl:N	**Conventional**	**3:1 Boost**	**5:1 Boost**
3:1	0.37	0.29	0.24
5:1	0.13	0.07	0.06

Figures 4.49 and 4.50, and 4.51 and 4.52 show the chloramine decay and ammonia formation with an initial Cl:N of 3:1 and 5:1, respectively, and boost doses to a Cl:N of 3:1 (Figures 4.49 and 4.51) or 5:1 (Figures 4.50 and 4.52). While the initial boost scenario doses applied less disinfectant mass than the conventional doses, the difference in the doses are very little which result in little differences between the associated chloramine decay and ammonia concentration. Additionally, an initial Cl:N of 5:1 resulted in a third of the ammonia as an initial Cl:N of 3:1 indicating the initial dose may significantly impact the reduction of ammonia concentration. The various booster disinfection scenarios were all able to reduce the ammonia concentration after the chlorine boost. For an initial Cl:N of 3:1, boosting to a Cl:N of 5:1 resulted in 50% less ammonia than boosting to a Cl:N of 3:1 due to the larger amount of chlorine available for recombining with the ammonia. For an initial Cl:N of 5:1, there was little observed difference in ammonia formation when boosting to a Cl:N of 5:1 or 3:1. The lack of observable difference may be related to the relatively low ammonia concentration present before the boost such that differences in ammonia concentration may be difficult to determine. Table 4.7 provides the average ammonia concentration from the different disinfection scenarios with a chloramine initial dose followed by a chlorine boost dose.

For this particular water, applying chloramine at an initial Cl:N of 5:1 followed by a boost dose to a Cl:N of 3:1 or 5:1 appeared to be the best disinfectant strategy for reducing the average ammonia concentration. Additionally, this scenario avoids the large chlorine doses for boosting needed under an initial Cl:N of 3:1 to recombine the ammonia present at the time of boosting.

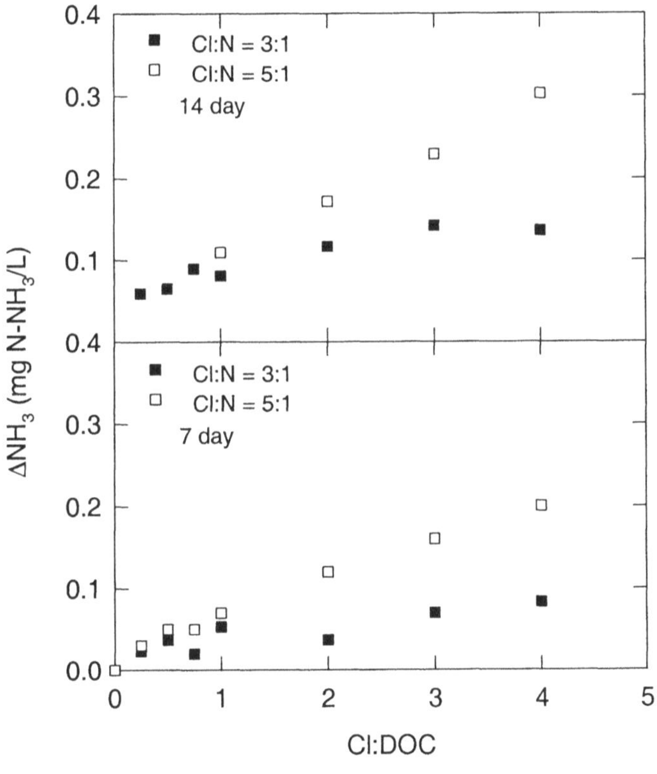

Figure 4.48: Change in ammonia concentration for Cl:N of 3:1 and 5:1 after the 7 and 14 day demand studies.

4.6.3 Summary

The use of booster disinfection did not significantly reduce the amount of ammonia required under the experimental conditions studied. A more substantial reduction in the pre-boost ammonia concentration occurs when using a Cl:N of 5:1 instead of 3:1 due to the increased chlorine concentration available to combine with the ammonia. Booster chlorination significantly reduced the ammonia concentration by recombining with the ammonia remaining in the water from the initial dose or resulting from chloramine decay. For an initial Cl:N of 3:1, booster chlorination to achieve a Cl:N of 5:1 significantly reduced the ammonia concentration relative to boosting to a Cl:N of 3:1 which, again, was due to the increased amount of chlorine available to recombine with the ammonia. For an initial Cl:N of 5:1, the reduction in ammonia concentration resulting from booster chlorination appeared independent of the final Cl:N, however, the ammonia concentrations before boosting were relatively low, potentially preventing a distinction to be made between the ammonia concentration resulting from a booster chlorination to a Cl:N of 3:1 or 5:1.

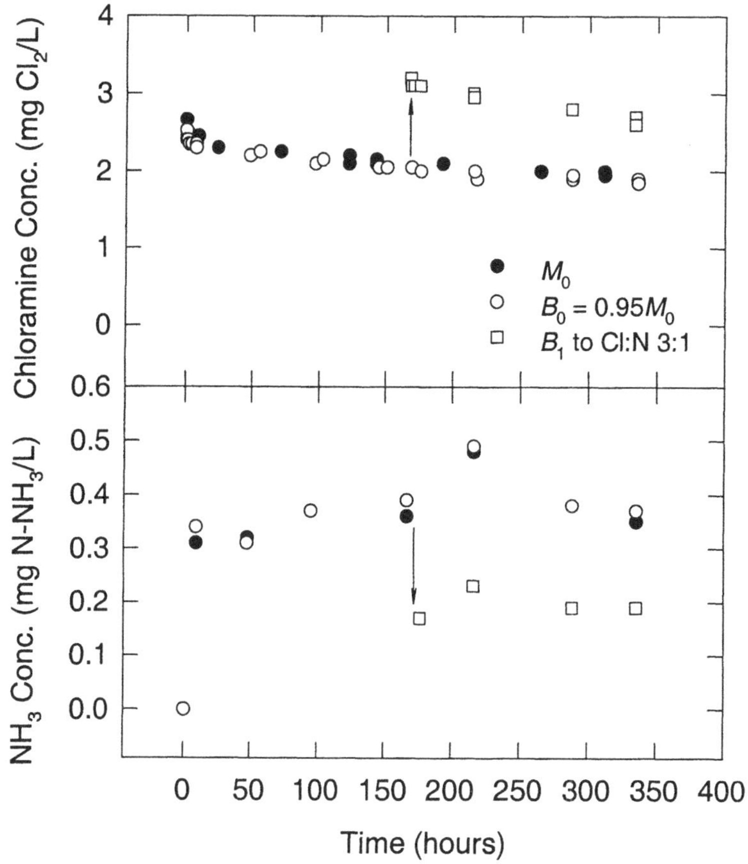

Figure 4.49: Chloramine decay and ammonia concentration from the Portland decay experiments with an initial and boost Cl:N of 3:1.

4.7 A Reactive Species Model

Relatively little research has developed models of chlorine loss and THM formation in natural waters under conditions where the chlorine dose changes (and, as a result, where the background matrix also changes). In addition to water treatment processes, such changes may occur via booster disinfection, where the reactive material characteristics at the time of rechlorination would be different from those at the time of the first dose. In addition to rechlorination at engineered booster stations, rechlorination also occurs in distribution systems through the mixing of waters with different average residence times. Such mixing is common since water that mixes at pipe junctions may follow different paths, resulting in different chlorine concentrations and reactive species characteristics. The chlorine decay kinetics after mixing (or, rechlorination), are not explicitly represented in typical chlorine decay models. Those models which could, in principle, be extended to represent chlorine decay kinetics after rechlorination (i.e. those including a reactive component), have not been evaluated using data from rechlorination experiments.

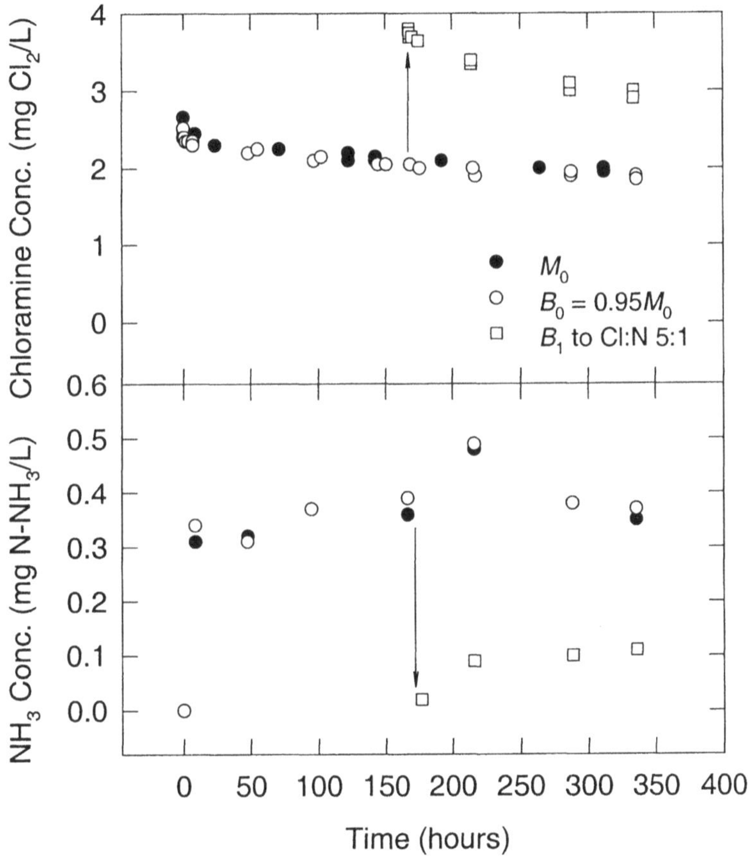

Figure 4.50: Chloramine decay and ammonia concentration from the Portland decay experiments with an initial Cl:N of 3:1 and boost Cl:N of 5:1.

This section extends the second–order, two–component model investigated by Clark(Clark), to provide an analytical solution capable of describing chlorine decay and THM formation under single, and multiple, rechlorination scenarios. Such a model is applicable to describe chlorine decay after booster disinfection(Tryby et al., 1999), and may improve distribution system models where the mixing of different residence time water leads to different chlorine decay kinetics after mixing. The second–order two–component model results are compared against the more typical first-order model.

4.7.1 Model Development

Chlorine Decay

We assume that initial rapid decay occurs within the clear well or before the first consumer, and prediction of chlorine decay is required after these rapid reactions are complete. The

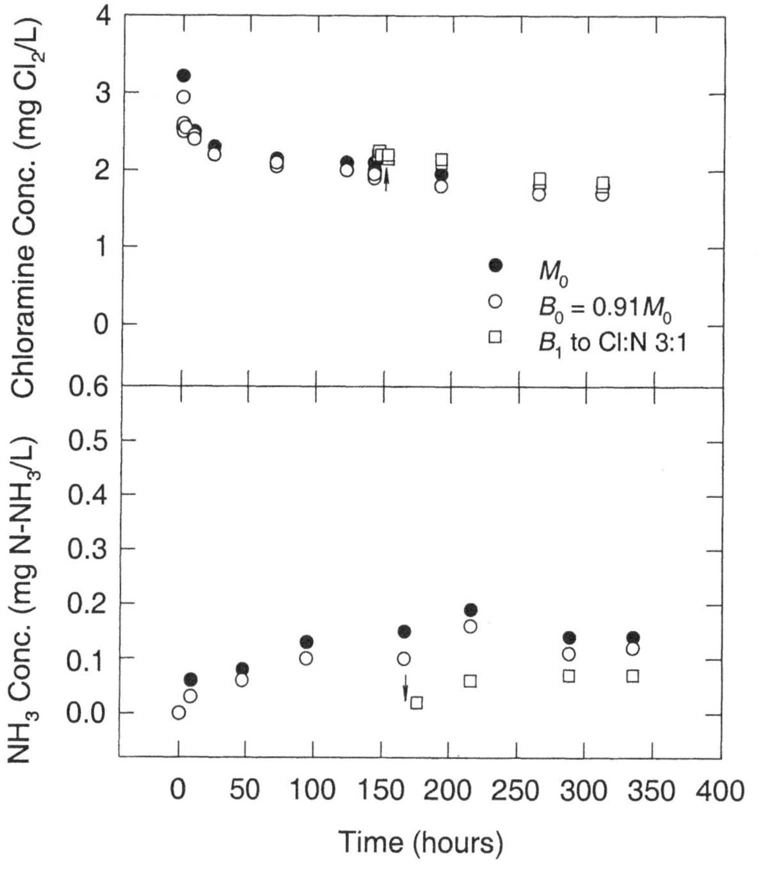

Figure 4.51: Chloramine decay and ammonia concentration from the Portland decay experiments with an initial Cl:N of 5:1 and boost Cl:N of 3:1.

solution to the first-order kinetic model in a batch system,

$$C(t) = C_0 e^{-kt}, \tag{4.15}$$

has been used extensively to predict free chlorine loss, where $C(t)$ is the disinfectant concentration (M/L^3) at time t, $C_0 = C(t = 0)$ is the initial concentration (M/L^3) (e.g. leaving the clearwell or before the first consumer), k is the first-order decay rate coefficient $(1/T)$, and t is the reaction time (T) (sometimes approximated by the average residence time, or water *age*, to a location in a distribution system). The first-order model inherently assumes that reactive material is always in excess. This assumption allows the disinfectant decay rate to depend only on the disinfectant concentration, and be independent of the observed chlorine demand (chlorine demand being an indicator of oxidized reactive material). Boccelli et al.(Boccelli et al., 1997) rechlorinated a natural water to the same chlorine concentration at different reaction times, and results showed the excess reactant assumption can be violated under conditions seen in field practice. As rechlorination times (and chlorine demand) increased, the estimated value of k decreased consistently, reflecting a limitation of the reactive material remaining in solution.

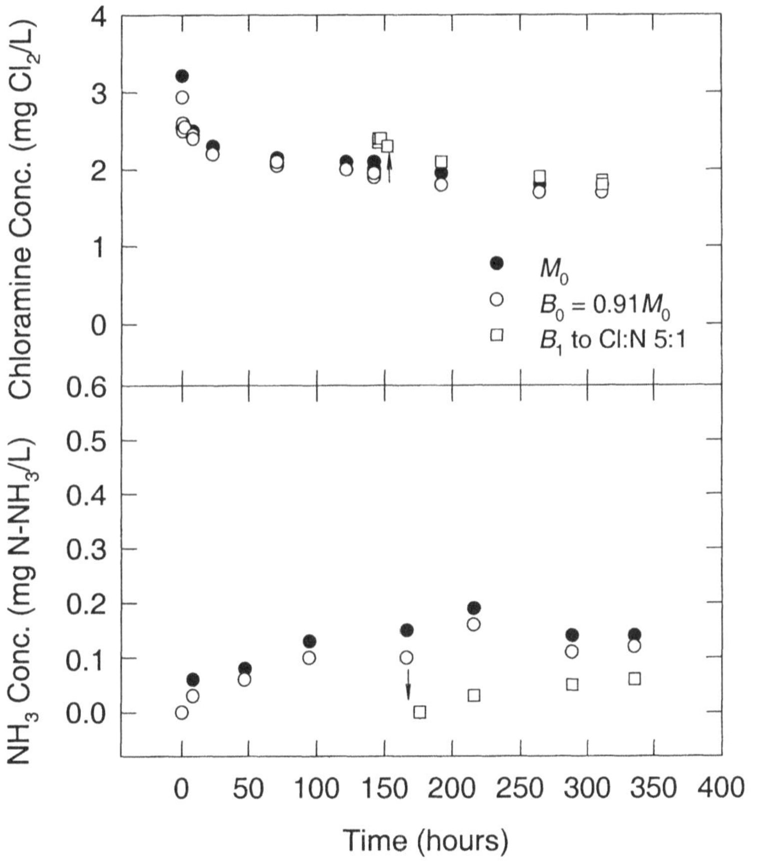

Figure 4.52: Chloramine decay and ammonia concentration from the Portland decay experiments with an initial and boost Cl:N of 5:1.

The second–order overall, two–component model investigated by Clark(Clark) accounts for both disinfectant and a fictitious reactant via the hypothetical irreversible reaction

$$\mathrm{aA} + \mathrm{bB} \rightarrow \mathrm{pP}, \tag{4.16}$$

where A is the free chlorine concentration, B is the fictive reactive component concentration, P is the disinfectant by-product concentration, and a, b, and p are stoichiometric reaction coefficients. This hypothetical reaction does not represent the actual mechanisms, but is the simplest one that is consistent with a conceptual model of free chlorine oxidizing natural organic matter. Clark(Clark) assumed the reaction rate to be first–order with respect to A and B, and second–order overall,

$$\frac{dC_A}{dt} = -k_A C_A C_B, \quad \frac{dC_B}{dt} = -k_B C_A C_B, \tag{4.17}$$

where C and k are the concentration and decay rate coefficient for the disinfectant (subscript A) and reactive component (subscript B) respectively, and $k_B = (b/a)k_A$. The analytical solution to (4.17) is(Levenspiel, 1972),

$$C_A(t) = \frac{C_{A,0} - \frac{aC_{B,0}}{b}}{1 - \frac{aC_{B,0}}{bC_{A,0}} \exp[-(\frac{bC_{A,0}}{aC_{B,0}} - 1)k_A C_{B,0} t]}, \tag{4.18}$$

where $C_{A,0}$ and $C_{B,0}$ are the initial concentrations of the disinfectant and fictive reactive component at $t = 0$, and a, b, $C_{B,0}$, and k_A are parameters that must be estimated from experimental data.

Clark and Sivaganesan(Clark and Sivaganesan, 1998) reduced the number of parameters in (4.18) to two in the equivalent solution,

$$C_A(t) = \frac{C_{A,0}(1-K)}{1 - K\exp(-ut)}, \tag{4.19}$$

in which

$$u = M(1-K), \quad K = \frac{aC_{B,0}}{bC_{A,0}}, \quad M = \frac{k_A b C_{A,0}}{a}, \tag{4.20}$$

where K (dimensionless) and M $(1/T)$ are estimated through non-linear empirical regression models that depend on natural water characteristics. The regression models used to estimate K and M were inconsistent with (4.20) in that K was not linear in $1/C_{A,0}$ and M was independent of $C_{A,0}$.

We consider a form of (4.18) with two parameter groups that are independent of $C_{A,0}$,

$$C_A(t) = \frac{C_{A,0} - \alpha}{1 - \frac{\alpha}{C_{A,0}} \exp[-(\frac{C_{A,0}}{\alpha} - 1)\beta t]}, \tag{4.21}$$

where

$$\alpha = \frac{aC_{B,0}}{b}, \quad \beta = k_A C_{B,0}. \tag{4.22}$$

The parameter α (M/L^3) is the "stoichiometric" chlorine concentration required for reaction (4.16) to go to completion, and β $(1/T)$ is a pseudo first–order decay rate coefficient. Since α and β are independent of $C_{A,0}$ (a measurable quantity), the model can be evaluated consistently for a given set of model parameters, and a range of initial conditions. Also, as discussed below, this form of the model can be extended conveniently to rechlorination scenarios.

For application to rechlorination conditions, (4.21) must obey the discontinuous boundary condition associated with a step input of disinfectant concentration. Thus, a separate model is required to describe the initial dose and each subsequent rechlorination dose. The disinfectant concentration after the initial dose is $C_A = f(C_{A,0}, \alpha, \beta, t)$, $0 \le t < t_b$, and the concentration after rechlorination is $C_A^* = f(C_{A,0}^*, \alpha^*, \beta^*, t^*)$, $t^* \ge 0$, where $t^* = t - t_b$, $C_{A,0}^*$ is the concentration immediately after rechlorination, α^*, β^* are model parameters after rechlorination (see below), and t_b is the time of rechlorination. Figure 4.53 illustrates the use of individual models in series to describe rechlorination conditions.

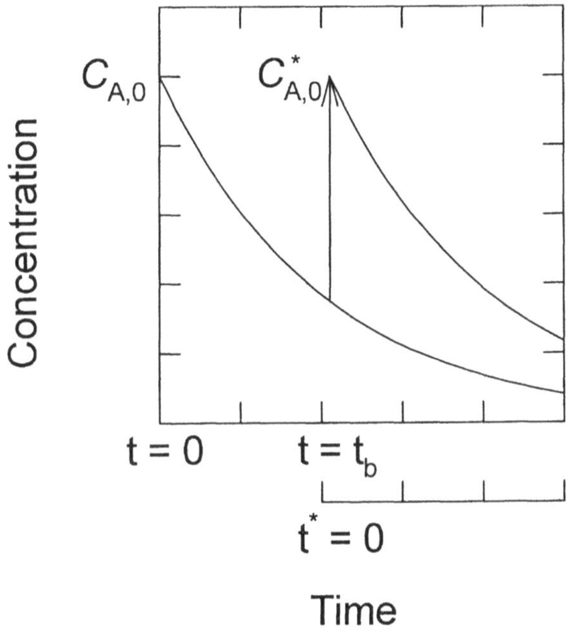

Figure 4.53: A series of separate models used to describe rechlorination kinetics.

The parameters α and β depend on stoicheometric and kinetic coefficients a, b, and k_A; these coefficients should be independent of chlorine concentration, dose history, and reaction time. The parameters α and β also depend, however, on the initial concentration of reactive species B for each separate model in series; this initial concentration – and thus the model parameters – is altered by reaction with free chlorine, in a manner that is consistent with kinetic model assumptions,

$$C^*_{B,0} = C_{B,0} - \frac{b}{a}[C_{A,0} - C_A(t_b)], \tag{4.23}$$

where $C_{A,0} - C_A(t_b)$ is the previous dose chlorine demand at the time of rechlorination, and b/a is the stoichiometric ratio of reactive material consumed to free chlorine consumed. The model parameters α and β are thus updated after rechlorination,

$$\alpha^* = \frac{aC^*_{B,0}}{b}, \quad \beta^* = k_A C^*_{B,0}, \tag{4.24}$$

and thus, letting $x(t_b) = C_{A,0} - C_A(t_b)$,

$$\alpha^* = \frac{a}{b}C_{B,0} - x(t_b) = \alpha - x(t_b), \tag{4.25}$$

$$\beta^* = k_A C_{B,0} - k_A \frac{b}{a} x(t_b) = \beta(1 - \frac{x(t_b)}{\alpha}). \tag{4.26}$$

Consistent with model assumptions, the post-rechlorination parameters α^* and β^* are defined as functions of the pre-rechlorination chlorine demand and model parameters α and β. Since

$x(t_b)$ depends on α and β, the models before and after rechlorination are one and the same, dependent only on the two parameters α and β, and on the measurable initial chlorine concentrations for each dose.

Model Exploration

In this section, we clarify the roles of α and β in the chlorine dynamics. The analysis is presented for the basic model depending on α and β but also apply to the rechlorination model depending on $C^*_{A,0}$, α^*, and β^*. The limiting chlorine concentration as $t \to \infty$ depends on the ratio $C_{A,0}/\alpha$, which determines the signs of the numerator and the exponential in (4.21). Recalling that α is the stoicheometric chlorine concentration for (4.16) to go to completion, the case $C_{A,0}/\alpha > 1$ corresponds to excess chlorine compared to reactive species, and α is the ultimate chlorine demand,

$$\lim_{t\to\infty} C_A(t) = \frac{C_{A,0} - \alpha}{1 - 0} = C_{A,0} - \alpha, \tag{4.27}$$

while if $C_{A,0}/\alpha \leq 1$ then reactive species is in excess compared to chlorine, and

$$\lim_{t\to\infty} C_A(t) = \frac{C_{A,0} - \alpha}{1 - \infty} = 0. \tag{4.28}$$

The parameter β affects only the rate at which free chlorine concentration converges to these limiting values $C_{A,0} - \alpha$ ($C_{A,0}/\alpha > 1$) or 0 ($C_{A,0}/\alpha \leq 1$).

The chlorine concentration in the limit when $\alpha \to \infty$ illustrates a connection with the typical first-order kinetic model ($\beta \to \infty$ simply implies an infinite decay rate, and the chlorine concentration in this case is equivalent to that when $t \to \infty$). Evaluating (4.21) term by term,

$$\lim_{\alpha\to\infty} (C_{A,0} - \alpha) = \lim_{\alpha\to\infty} -\alpha, \tag{4.29}$$

$$\lim_{\alpha\to\infty} (1 - \frac{\alpha}{C_{A,0}} \exp[\ldots]) = \lim_{\alpha\to\infty} (-\frac{\alpha}{C_{A,0}} \exp[\ldots]), \tag{4.30}$$

$$\lim_{\alpha\to\infty} \exp[-(\frac{C_{A,0}}{\alpha} - 1)\beta t] = \exp[-(0-1)\beta t] = \exp[\beta t], \tag{4.31}$$

and substituting (4.29-4.31) into (4.21) as $\alpha \to \infty$ yields

$$\lim_{\alpha\to\infty} C_A(t) = \lim_{\alpha\to\infty} \frac{-\alpha}{-\frac{\alpha}{C_{A,0}} e^{\beta t}} = C_{A,0} e^{-\beta t}. \tag{4.32}$$

Thus the second- and first-order models are equivalent in the limit when $\alpha \to \infty$, where $k = \beta$. This equivalence is sensible as α is the stoicheometric chlorine concentration required for reaction with $C_{B,0}$; large values of α imply that B is in significant excess ($C_{B,0}$ is large), or that relatively small quantities of reactive species are consumed per unit of chlorine consumed

(a/b is large); in either case one may consider $C_B(t) \approx C_{B,0}$, and thus use a single-species, first-order model where $k = k_A C_{B,0} = \beta$ (see 4.17). Further, the values of α and β are not affected by rechlorination in the limit $\alpha \to \infty$, since (4.25) and (4.26) show $\lim_{\alpha\to\infty} \alpha^* = \alpha$ and $\lim_{\alpha\to\infty} \beta^* = \beta$. Since the limiting behavior of the second-order model, under single dose or rechlorination conditions, is that of a first-order model with constant decay rate coefficient, the second-order model must represent chlorine decay data at least as well as a first-order model under single or multiple doses.

THM Formation

Clark(Clark,) noted that (4.16) implies chlorination by-products, such as THMs, are formed as a linear function of the free chlorine demanded,

$$THM(t) = T \cdot x(t) + M \tag{4.33}$$

where $x(t)$ is the total chlorine demand at time t, $T(= p/a)$ is a dimensionless parameter relating THM formation to chlorine demand, and M is the THM concentration present at $t = 0$ ($M > 0$ if the model is applied to waters that were previously chlorinated).

The THM model (4.33) is applied to single- or multiple-dose scenarios, since model assumptions imply that THMs are always formed in the same proportion (p/a) to chlorine consumed (see 4.16). Of course, the chlorine demand under rechlorination scenarios must be calculated as described above for each dose, and then summed to yield the total demand $x(t)$.

4.7.2 Results and Discussion: Chlorine Decay

Five different waters, listed in Table 4.1, were used to evaluate rechlorination kinetics. The "Labels" in Table 4.1 will be used to reference the five waters throughout the modeling discussion.

Model Start Time

Typical chlorine decay models are unable to accurately describe the initial, rapid chlorine demand typically associated with natural waters, and we investigate the ability of the second-order model to capture this behavior. Figure 4.54 illustrates calibrated first- and second-order models for the TX water by starting the models 1, 2, 4, or 8 hours after the initial dose. The parameter values (α and β, or k) minimize the squared errors between model predictions and measurements, and the initial chlorine concentration ($C_{A,0}$) is taken as the measured value at the respective start time. These results represent those observed for all other waters.

The results suggest neither model is able to accurately describe the entire range of chlorine decay behavior starting with an initial dose (results for a start time immediately after the

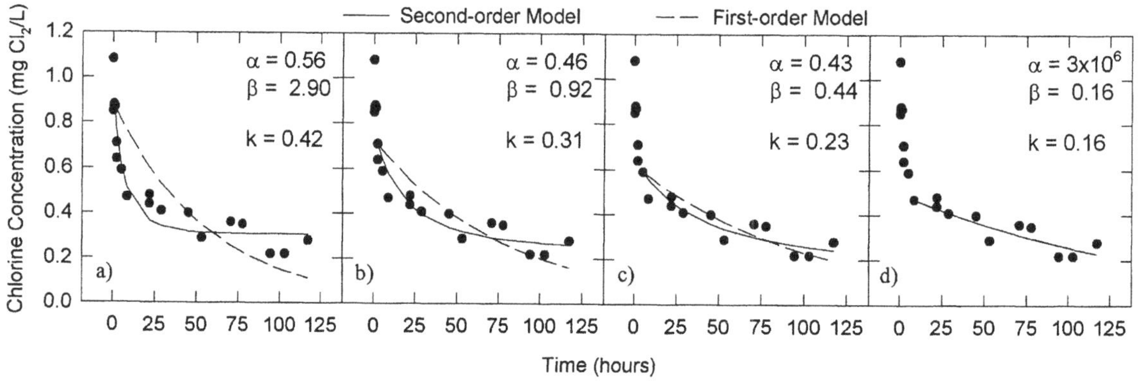

Figure 4.54: Calibrated first- and second-order models using TX data and varying start time. Models started 1 hour (a), 2 hours (b), 4 hours (c), and 8 hours (d) after initial dose.

initial dose are not shown, but are consistent with the trend shown in figure 4.54). The accuracy of both models improved significantly as the model start time increased, thus ignoring the initial rapid decay. At later model start times, first- and second-order models were equivalent for three of five waters (OH, TX, CT), as measured by the root mean square error (RMSE). For the TX water (Figure 4.54), the first- and second-order models are equivalent if started 8 hr. after the initial dose, when the associated second-order parameter α becomes large ($10^5 - -10^6$) (and thus the models become equivalent, see 4.32).

Comparison of First- and Second-Order Models

First- and second-order model parameters (k, or α and β) were estimated for all waters listed in Table 4.1, using data from the initial dose plus n_b rechlorination doses. For example, the HLW water experiments (for a given Cl:DOC) used a single initial dose followed by nine different rechlorination doses ($n_b = 9$); the model parameters are estimated by minimizing the sum of squared errors for the initial and nine rechlorination doses simultaneously (a total of 10 different doses). All calibration results use a model start time 4 hours after the initial dose, and measured values of the initial concentrations $C_{A,0}$ and $C^*_{A,0}$. The second-order model is estimated using (4.21) to model the initial dose, and (4.25) and (4.26) to calculate parameters α^* and β^* for each rechlorination dose. The first-order model is estimated by simply identifying the single decay rate constant, k, that best represents the initial and all

Table 4.8: First- and second-order model parameter estimates and "goodness-of-fit" measurements for all waters, incorporating both initial and all boost doses

					Second-Order			First-Order	
Water	Initial Dose (mg/L)	n_b	Boost Conc. (mg/L)	Boost Time (hrs.)	α	β	RMSE (mg/L)	k (d^{-1})	RMSE (mg/L)
HLW	8.75	9	4.83-8.75	24-430	7.62	0.20	0.39	0.09	0.73
	7.00	9	4.26-7.00	24-310	5.19	0.28	0.24	0.11	0.60
	5.25	9	2.50-5.25	24-168	4.92	0.33	0.19	0.14	0.36
CT	1.61	1	0.71	72	1.48	0.94	0.05	0.53	0.08
	1.07	1	1.18	24	1.40	1.02	0.06	0.51	0.09
NV	1.80	1	0.62	72	1.47	0.83	0.03	0.52	0.06
	1.35	1	1.07	24	1.19	1.09	0.05	0.45	0.09
OH	0.91	1	0.49	72	0.57	0.36	0.03	0.22	0.04
	0.71	1	0.65	24	0.61	0.36	0.03	0.23	0.04
TX	0.98	1	0.37	72	0.38	0.30	0.04	0.19	0.04
	0.88	1	0.51	24	0.35	0.57	0.04	0.21	0.06

rechlorination doses. Table 4.8 provides the parameter estimation results, and information regarding the initial concentration, number of rechlorination doses n_b, the range of rechlorination concentration and time, and a "goodness-of-fit" statistic (RMSE).

The results in Table 4.8 suggest that the differences between first- and second-order models may often be slight, although in some cases the second-order model is significantly better for describing chlorine decay under rechlorination conditions. Like the first-order model, the optimal second-order parameters appear to depend on initial and rechlorination dose magnitude and on rechlorination time; one should not consider second-order model parameters as "universal" for a given water, but rather experiments should reflect expected dose conditions. The value of α does not become large under rechlorination conditions (in contrast to when only initial dose data are used). The assumption of a constant "reactive" component is less likely to be valid under rechlorination scenarios, and thus the first-order model is less likely to adequately represent chlorine decay under those conditions (Boccelli et al.(Boccelli et al., 1997) showed the individual dose first-order decay coefficient decreased after rechlorination, which also supports a reduction in "reactive" material concentration).

Figure 4.55 shows representative calibrated first- and second-order models for the NV and OH waters (complete model results are provided in (Boccelli, 1999)). Visual inspection of Figure 4.55 illustrates the improved performance of the second-order model over the first-order model. The first-order model, with a constant decay coefficient, can not account for changes in the "reactive" component, and is forced to compromise between larger decay rate constants that are indicated at early times, and relatively smaller decay rate constants at later times.

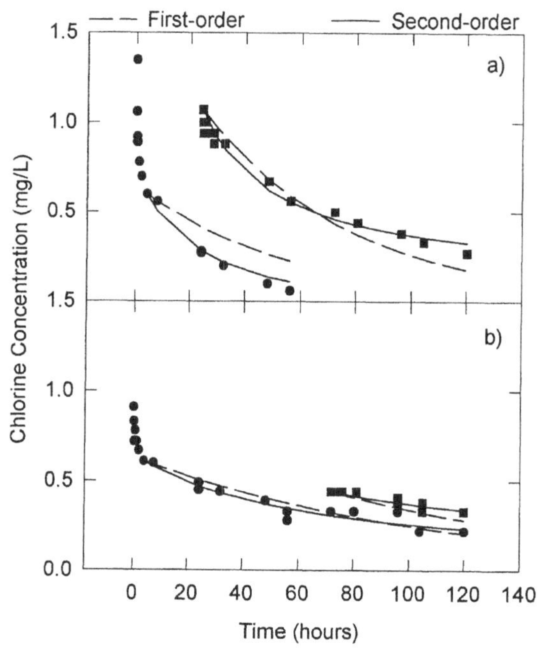

Figure 4.55: Calibrated first- and second-order models, using all initial and rechlorination data: a) NV water with initial dose of 1.35 mg/L and rechlorination after 24 hours; b) OH water with initial dose of 0.91 mg/L and rechlorination after 72 hours. (round symbols = initial dose; square symbols = rechlorination dose)

While differences between first- and second-order models may often be small, larger discrepancies will occur if chlorination conditions allow for significant changes in the reactive component. Intuitively, such conditions lean toward higher doses, or multiple doses, over longer reaction times. Previous studies that found little difference between first- and second-order models(Vasconcelos et al., 1996) did not evaluate them under rechlorination conditions, which is apparently where any discrepancy is most likely to be noticed. Under single-dose conditions, a large reactive component change is likely to coincide with a large decrease in free chlorine concentration, and thus a relatively small free chlorine decay rate – which tends to mask any errors in decay rate. Under rechlorination conditions, however, errors in decay rate caused by changes in the reactive component are amplified after rechlorination. In particular, under rechlorination there is the opportunity to observe decay rates at the same chlorine concentration but two different reaction times; a significant difference in these

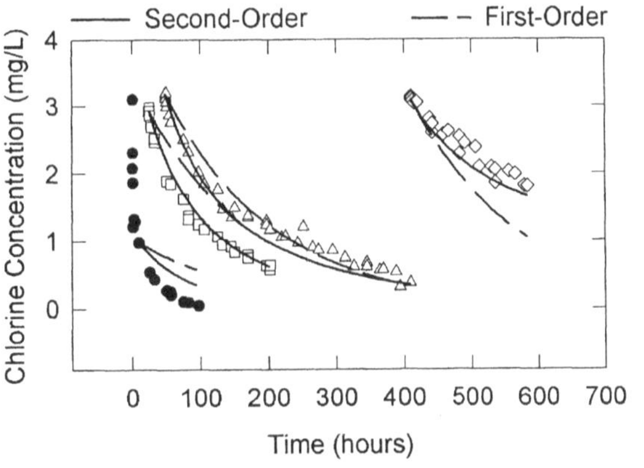

Figure 4.56: Calibrated first- and second-order models for multiple rechlorination experiments on HLW. Rechlorination to the initial dose concentration (3.10 mg/L) occured 24, 48, and 410 hours after initial dose. (round symbols = initial dose; other symbols = rechlorination doses)

two rates would point to an inconsistency of a first-order modeling approach. For example, figure 4.56 shows results from HLW experiments (not shown in table 4.8) in which the same water was rechlorinated three times in sequence. The inadequacy of a first-order model is apparent.

4.7.3 Results and Discussion: THM Formation

Linear Relationship of THM Formation and Chlorine Demand

The assumed linear relationship between THM formation and chlorine demand (4.33) was tested using measured chlorine and THM values. Figure 4.57 shows the calibrated linear models for CT, NV, OH, and TX waters, along with the linear correlation coefficients and estimated model parameters T and M. The parameter T represents the slope of each line in Figure 4.57, while M represents its intercept. Both visual inspection (Figure 4.57) and values of the correlation coefficient ρ indicate the proposed linear model adequately represents the relationship between THM formation and chlorine demand under rechlorination conditions.

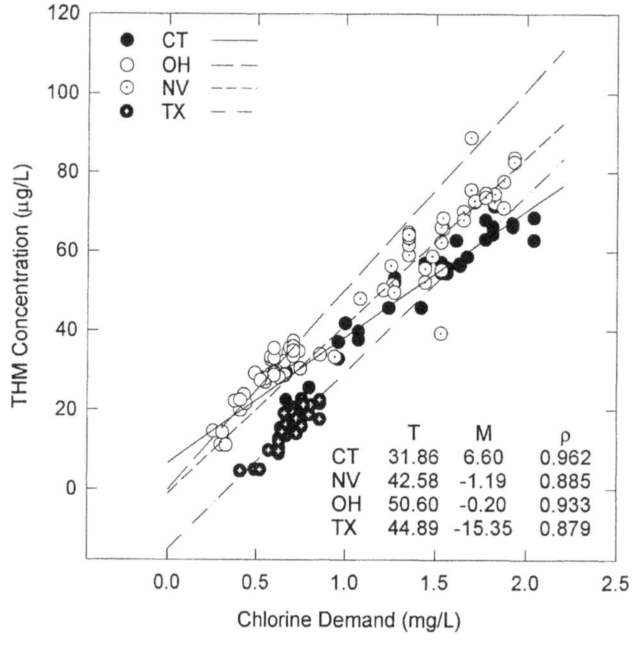

Figure 4.57: Estimated linear relationships between measured THM formation and chlorine demand.

The different values of T (Figure 4.57) are not surprising as THM formation from available organic matter is expected to depend on the source water and treatment processes. Since these waters had not been previously chlorinated, the estimates of M were expected to be negative to account for inorganic reactions that do not form THMs (such reactions are often assumed to be responsible for rapid decay of chlorine after the first dose). The only water with M statistically less than zero is TX – a ground water which would typically have high inorganic concentrations. The value of M for the NV and ORW waters are not statistically different from zero, suggesting that inorganic reactions have relatively little impact on chlorine demand. The positive value of M for the CT water suggests there may have been a small amount of THMs present prior to dosing, although this was not confirmed by analysis since THMs were not suspected in unchlorinated source water. In all cases, however, the linear THM formation model adequately represents THM formation as a function of measured chlorine demand.

Table 4.9: THM model performance for initial dose and rechlorination data, using chlorine demand from second-order model. THM parameter estimates obtained using initial and rechlorination data, and either calculated (second-order) or measured chlorine demand.

				Calculated Chlorine Demand			Experimental Chlorine Demand		
Water	Initial Dose (mg/L)	Boost Conc. (mg/L)	Boost Time (hrs.)	T ($\mu g/mg$)	M ($\mu g/L$)	RMSE ($\mu g/L$)	T ($\mu g/mg$)	M ($\mu g/L$)	RMSE ($\mu g/L$)
CT	1.61	0.71	72	31.65	29.27	1.63	29.04	30.53	1.97
NV	1.07	1.18	24	34.40	26.00	3.94	34.36	24.75	4.14
	1.80	0.62	72	43.51	34.76	5.76	46.15	31.82	5.88
OH	0.91	0.49	72	55.96	12.43	1.28	51.13	14.37	1.39
	0.71	0.65	24	48.21	16.83	3.03	51.09	16.86	3.12
TX	0.98	0.37	72	58.13	5.61	1.62	59.03	6.36	1.89
	0.88	0.51	24	49.83	2.77	1.66	37.13	4.12	2.47

THM Formation Modeling

One cannot always use measured chlorine demand to predict THM formation; in practice it would often be desirable to use model predicted chlorine demand to predict THM formation. Here the THM formation model using model predicted chlorine demand is evaluated using parameters, T and M, that are estimated two ways: 1) using chlorine demand data (as above), and 2) using chlorine demand estimated by the second-order chlorine decay model. In general, the THM model coupled with the second-order chlorine decay model produced better THM estimates than when coupled with the first-order chlorine decay model, due to more accurate chlorine demand estimation(Boccelli, 1999).

THM formation model parameters were estimated using an initial dose (with a model start time of 4 hours) and n_b rechlorination doses. This was done for all waters listed in Table 4.1, with the exception of HLW. Table 4.9 presents THM model parameter estimates using calculated (second-order model), and measured, chlorine demand. The THM model parameter estimates obtained using the measured demand compared favorably with those obtained using the calculated demand. The "experimental chlorine demand" parameter values in Table 4.9 differ from those in figure 4.57 because the model is started 4 hours after the initial dose, and calibrated models are obtained for each different initial and rechlorination dose; thus, for example, the value of M in Table 4.9 represents the THM concentration 4 hours after the initial dose.

Figure 4.58 presents representative THM model results for the NV and TX waters, with parameters determined using experimental and calculated chlorine demand data. The THM formation calculated using model parameters estimated from measured chlorine demand data is comparable that calculated using calculated chlorine demand. The chlorine demand

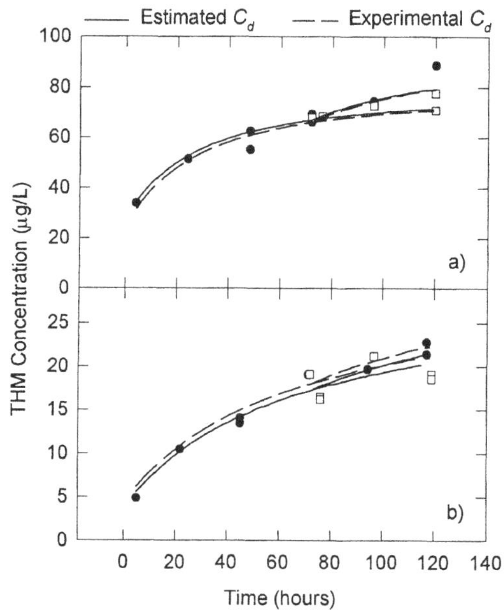

Figure 4.58: THM model predictions for a) 72 hour rechlorination of NV water, and b) 72 hour rechlorination of TX water. THM model parameters estimated using both calcualted and experimental chlorine demand. (round symbols = initial dose; square symbols = rechlorination dose)

calculated using the second-order model allowed for reliable estimation of THM formation using a linear relationship. Further, to the extent that THM formation and chlorine demand are linearly related, the demand calculated by a second-order model should be preferred to a first-order model, whenever changes in the reactive components lead to significant deviations from first-order chlorine decay behavior.

4.7.4 Water Quality Modeling Implications

The above results illustrate that it can be important to include a reactive species when modeling chlorine decay under rechlorination conditions. Rechlorination conditions occur in practice when booster chlorination is practiced in the distribution system(Tryby et al., 1999). Rechlorination conditions also occur in distribution systems, however, even if booster chlorination is not employed. Distribution systems are looped networks, allowing for mixture of water parcels travelling along different paths with different reaction times, and thus creating

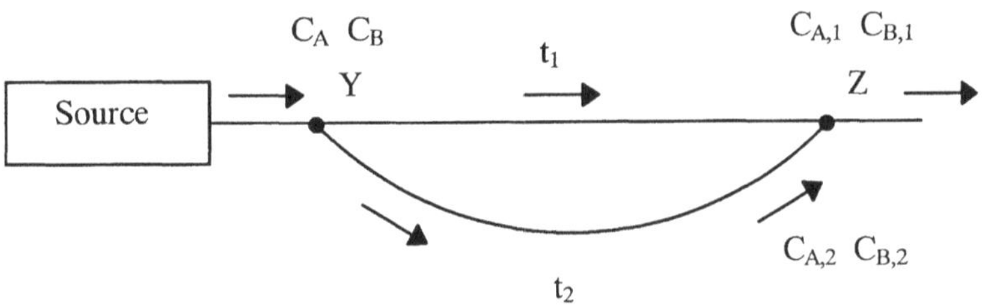

Figure 4.59: Simple distribution system showing the mixing of water with different reaction times.

an effect similar to booster chlorination; the network in figure 4.59 provides a simple illustration of this common situation. Water leaving the source has a disinfectant concentration C_A and reactive component concentration C_B at point Y, where the flow branches off into two different paths. A fraction of the source water requires a time t_1 to reach point Z, while the remaining water requires a time t_2 to reach the same point. Assuming $t_1 \ll t_2$, C_A and C_B entering point Z with travel time t_1 will be significantly larger than the concentrations with travel time t_2. Therefore, the disinfectant concentration from path 2 is being "boosted" (or increased) by the disinfectant concentration entering from path 1.

A second-order model of chlorine decay with a reactive constituent can be incorporated in distribution network water quality models, but not without important changes to the structure of typical network model algorithms. These changes are required since typical algorithms only model the dynamics of a single water quality constituent (e.g., free chlorine). While it is true the system (4.17) can be written equivalently in terms of a single second-order kinetic model for free chlorine, using $C_B(t) = C_{B,0} - (b/a)(C_{A,0} - C_A(t))$,

$$\frac{dC_A}{dt} = k_1 C_A + k_2 C_A^2, \tag{4.34}$$

where,

$$k_1 = \frac{b}{a} k_A C_{A,0} - k_A C_{B,0}, \tag{4.35}$$

$$k_2 = -k_A \frac{b}{a}, \tag{4.36}$$

such a kinetic model is ill-suited for use within a single-species network water quality algorithm. The difficulty is that the kinetic coefficient k_1 is not constant but rather depends on $C_{A,0}$ and $C_{B,0}$, which must change with every change in boundary condition (such as after the junction Z in figure 4.59). While it is possible to use the analytical solution (4.18) and compute new values of the model parameters after point Z in figure 4.59 (similar to what we have done in (4.25-4.26) for experimental data analysis), this approach would be of little use in general network algorithms that must use the general differential equation (4.34), rather than a particular solution to it. In particular, the use of a second-order kinetic expression such as (4.34) with *constant coefficients* is not equivalent to (4.17). The development of multi-component distribution system water quality models is required to model a reactive component in water distribution systems. Such a development could ultimately be used not only to model chlorine kinetics that depend on a reactive component but, by modeling the reactive component characteristics of each water source, to better model chlorine decay kinetics when different water sources are blended within the distribution system.

4.7.5 Reactive Species Model Conclusions

A second-order chlorine decay and THM formation model was investigated for rechlorination scenarios. The significance of such a model, relative to the typical first-order model, is the explicit consideration of a reactive species in the reaction kinetics. Such a formulation predicts that the chlorine decay rate, at a particular chlorine concentration, will decrease as the reaction continues – a feature which matches observations. In contrast, the first-order model assumes a constant chlorine decay rate at a particular chlorine concentration. The second-order model was extended to an arbitrary number of rechlorination events by showing how model parameters must change as a function of changes in the reactive constituent; this form of the model was used to interpret experimental rechlorination data. The second-order model was shown to always perform better than, or as well as, the first-order model.

The second-order chlorine decay model, like the first-order chlorine decay model (and most other decay models), was unable to adequately describe the initial rapid decay which generally occurred immediately after the initial chlorine dose. As a model for describing chlorine decay, the second-order model consistently outperformed the first-order model and provided a framework to incorporate a fictive reactive species. The use of a fictive reactive component allowed the second-order model to more accurately represent rechlorination kinetics than the first-order model. The magnitude of discrepancy between the two model descriptions are dependent both on the water being considered and the experimental conditions under which the chlorine is added to the water. Previous studies have shown little differences between first- and second-order models for chlorine decay, a fact that likely derives from the single-dose conditions of the experiments. Rechlorination conditions can amplify the errors associated with a first-order model, by exposing the differences in chlorine decay rates at identical chlorine concentrations, but different reaction times.

THM concentration was shown to be a strong linear function of the experimental chlorine demand. Estimating the THM formation model parameters by calculating the chlorine demand 1) using the second-order decay model, or 2) using the experimental data, both resulted in good representations of the observed experimental THM concentrations. Therefore, the THM model parameters can be determined directly from the data, rather than estimating them through a coupled chlorine decay and THM formation model.

The important feature of the presented second-order model is the consideration of a reactive species which decreases as the chlorine reactions proceed. Such a model can be generally applied to describe rechlorination kinetics. A relatively straightforward use of such a model would be for analyzing experimental data to simulate booster disinfection, or for multiple chlorine additions within a treatment process. Another application is to better model bulk chlorine decay (and perhaps THM formation) in distribution systems. Given that distribution systems are looped and branched networks, influent water to a particular junction node may have both differing chlorine and reactive species concentration. In effect, the water with the lower chlorine concentration is being rechlorinated, and the reactive species concentration is altered by mixing. This application would require network models to track the dynamics of both free chlorine and reactive component.

Chapter 5

Design of Booster Systems Using Network Models

Although booster disinfection is practiced in the United States and in Europe, the absence of a rigorous design method requires that stations be located and operated by trial and error, perhaps supported by simulation. Thus, booster stations are often located near areas experiencing obvious residual maintenance difficulties, and are operated according to local goals that ignore system-level interactions. For example, by strategically locating and operating a booster station, disinfectant dosages can be coordinated with those being made upstream to achieve a more desirable residual distribution throughout the entire network. The systems approaches presented here will identify such opportunities, as they simultaneously select multiple booster locations and schedule disinfectant mass dosage rates based on several competing objectives, subject to constraints that describe the dynamic transport and decay of disinfectant throughout the distribution system. The objectives considered include: 1) minimization of the total mass of disinfectant applied, 2) minimization of the total number of booster stations constructed, and 3) maintenance of adequate residual at specified locations and times.

The mathematical programming formulations presented here are based on the linear programming (LP) booster disinfection scheduling model presented by Boccelli et al.(Boccelli et al., 1998), which was developed by the authors for this AwwaRF project. The extensions allow the formulation to accommodate various disinfectant source models that are valuable for practical application of the designs produced, and incorporates booster station location as a decision variable within the optimization. In addition, a separate form of the model is developed as a maximum set covering model, which is amenable to solution via heuristic methods (as opposed to formal optimization methods). All the formulations presented here are coded for solution within the Booster Disinfection Design and Analysis (BDDA) software, described in Appendix A.

5.1 Minimum Mass Location and Scheduling Model

The minimum mass location and scheduling model is related to the general, mixed-integer linear programming (MILP), fixed-charge facility location problem(Efroymson and Ray, 1968), and is solved using a branch–and–bound solution procedure. The general fixed-charge facility location problem is well known in operations research. The following sections show how location and scheduling decisions for booster disinfection stations operating in water distribution networks can be expressed within the location problem framework. First, two common disinfectant source types, constant and flow proportional mass dosages, are described. An example using the two source types illustrates how linear superposition can be used to calculate the network response to disinfectant mass dosages. Next, the mathematics describing a generalized dosage parameterization that conforms with the assumption of linear superposition and allows flexible representation of source types within the booster location model is developed. Finally, the formulation of the booster location model is presented, and we discuss the conditions under which solutions are sensible for long-term planning and operation. Computational experience shows that the booster location model is a practical design method that can be applied to real networks (see chapter 6).

5.1.1 Disinfectant Source Parameterization

Meaningful analysis of water quality behavior within distribution systems requires that the variation in water demands, in pipe flows (rate and direction), and storage volumes be accounted for, due to their subsequent impact on disinfectant concentrations. Disinfectant sources and the strategies for regulating the dosages applied there can also affect concentrations. Several different strategies for dosage control are widely used - a constant dosage applied using a metering pump, or control based on a signal proportional to the flow rate past the injector. As we shall demonstrate in the following example, the type of source dosage control chosen influences the variability of disinfectant concentrations at the injector and throughout the network, and thus affects the practical implementation of booster disinfection station designs.

Consider the network shown in Figure 5.1. For planning purposes, the network is assumed to operate under a set of periodic nodal demands – a demand cycle which repeats indefinitely. Thus, an extended period hydraulic simulation(Rao and Bree, 1977) will typically yield (after a sufficient simulation time) a periodic set of network flows that also repeat indefinitely. In this particular case the demands and flow rates repeat on a 24 hour cycle. A disinfectant residual (e.g., free chlorine) is present in the network due to the addition of disinfectant mass at two locations, labeled "1" and "2" in the figure. First-order disinfectant decay kinetics are assumed(Rossman et al., 1994), and the residual at any location will vary over time in response to the variations in disinfectant injection rates and network hydraulics.

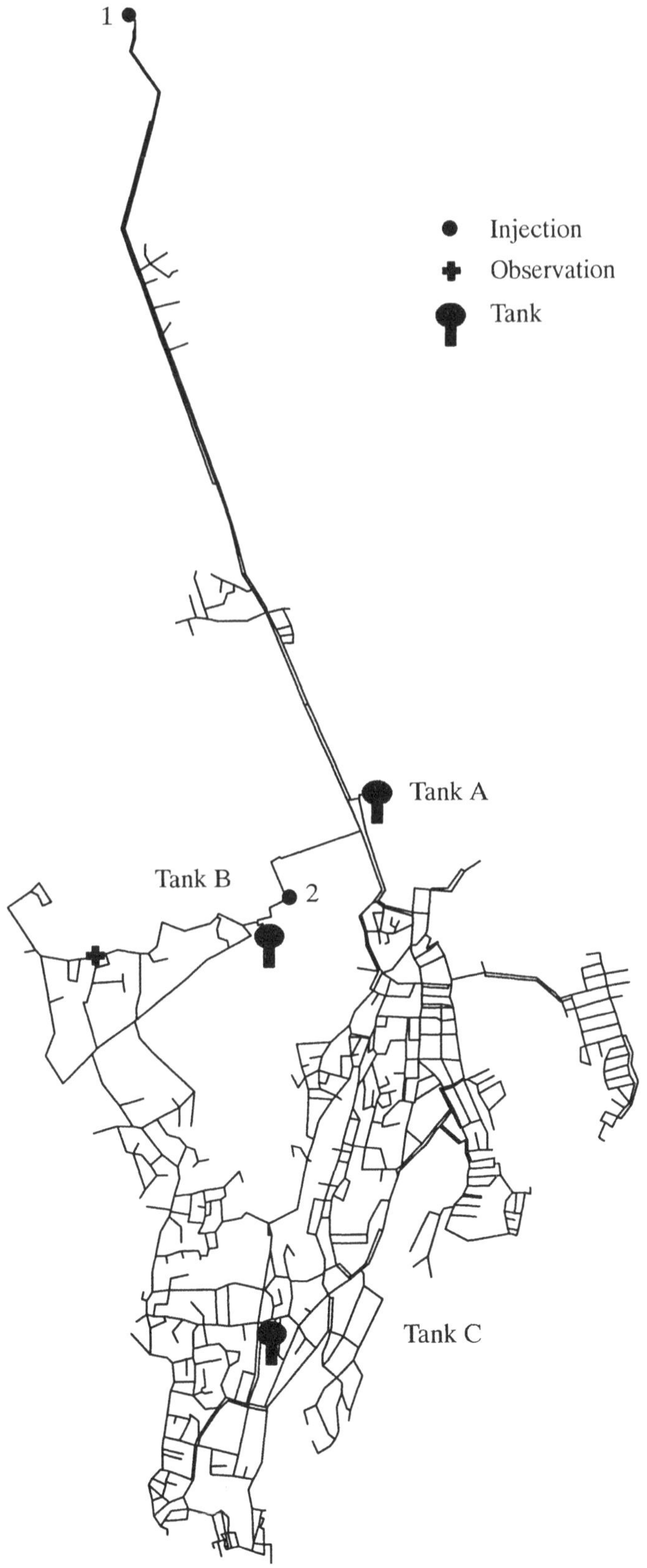

Figure 5.1: South Norwalk GFZ network map.

The disinfectant sources at injector locations 1 and 2 employ two different strategies for regulating dosages. Injector 1 delivers a flow proportional dosage (i.e. the mass dosage rate is time varying and proportional to the flow leaving the injector), while Injector 2 delivers a constant mass dosage rate. Figures 5.2(a) and (b) illustrate the dosage rate (shown in bold) and concentration input at each injector. Figure 5.2(a) illustrates the flow proportional dosage applied at Injector 1. The dosage rate varies dynamically, compensating for periodic changes in flow, and results in a uniform input concentration of 1.2 mg/l leaving the injector. Figure 5.2(b) illustrates the constant dosage applied at Injector 2. Unlike Injector 1, the input concentration at Injector 2 fluctuates significantly, as a result of periodic changes in flow and a constant dosage rate of 500 mg/min. Thus, flow proportional control like that at Injector 1 results in a constant input concentration and a periodic dosage rate, while a constant dosage rate like that at Injector 2 results in a periodic input concentration. This difference in source type behavior stems from the way changes in flow affect the dosage rate at the injector.

To understand these differences in source behavior more clearly, consider the disinfectant mass balance for a source location, generalized here as node i, which completely mixes its input flows with the disinfectant dosage applied at the injector,

$$c_i^*(t) = \frac{1}{Q_i(t)} \sum_{j \in IN_i(t)} q_j(t) c_j(t) + \frac{1}{Q_i(t)} v_i^k(t), \tag{5.1}$$

where $q_j(t)$ (L^3/T) is the flow in pipe j, $Q_i(t)$ (L^3/T) is the total flow into (and out of) node i, $c_j(t)$ (M/L^3) is the disinfectant concentration entering node i through pipe j, $IN_i(t)$ is a set of all pipe indices with flow entering node i, and $c_i^*(t)$ (M/L^3) is the completely mixed concentration assigned all flows leaving node i. The second term on the right side of (5.1) represents the concentration contribution due to the mass dosage rate $v_i^k(t)$ (M/T). The dosage rate, $v_i^k(t)$, is assumed for design purposes to be cyclical and to repeat indefinitely. For operational flexibility, the periodic operating cycle is discretized into n_i time intervals, each of length δt_i^k, $k = 1, \ldots, n_i$. Thus, within booster period k, the dosage is simply $v_i^k(t)$, and outside the interval $v_i^k(t) = 0$. For the constant dose case,

$$v_i^k = \gamma, \tag{5.2}$$

where γ (M/T) is a constant mass dosage rate applied at the injector. Constant dosage source types (like Injector 2) results in an input concentration equal to $\gamma/Q_i(t)$, which changes over time as a function of the total flow through the node. In water distribution systems, $Q_i(t)$ and thus $c_i^*(t)$ can vary significantly over a single day, because of diurnal variation in water usage rates. When an injector must accommodate variations in flow, one practical solution is a "flow proportional" controller, which uses flow rate measurements to adjust $v_i^k(t)$ in proportion to $Q_i(t)$. Thus, the dosage applied at the injector for the flow proportional case is,

$$v_i^k(t) = \beta Q_i(t), \tag{5.3}$$

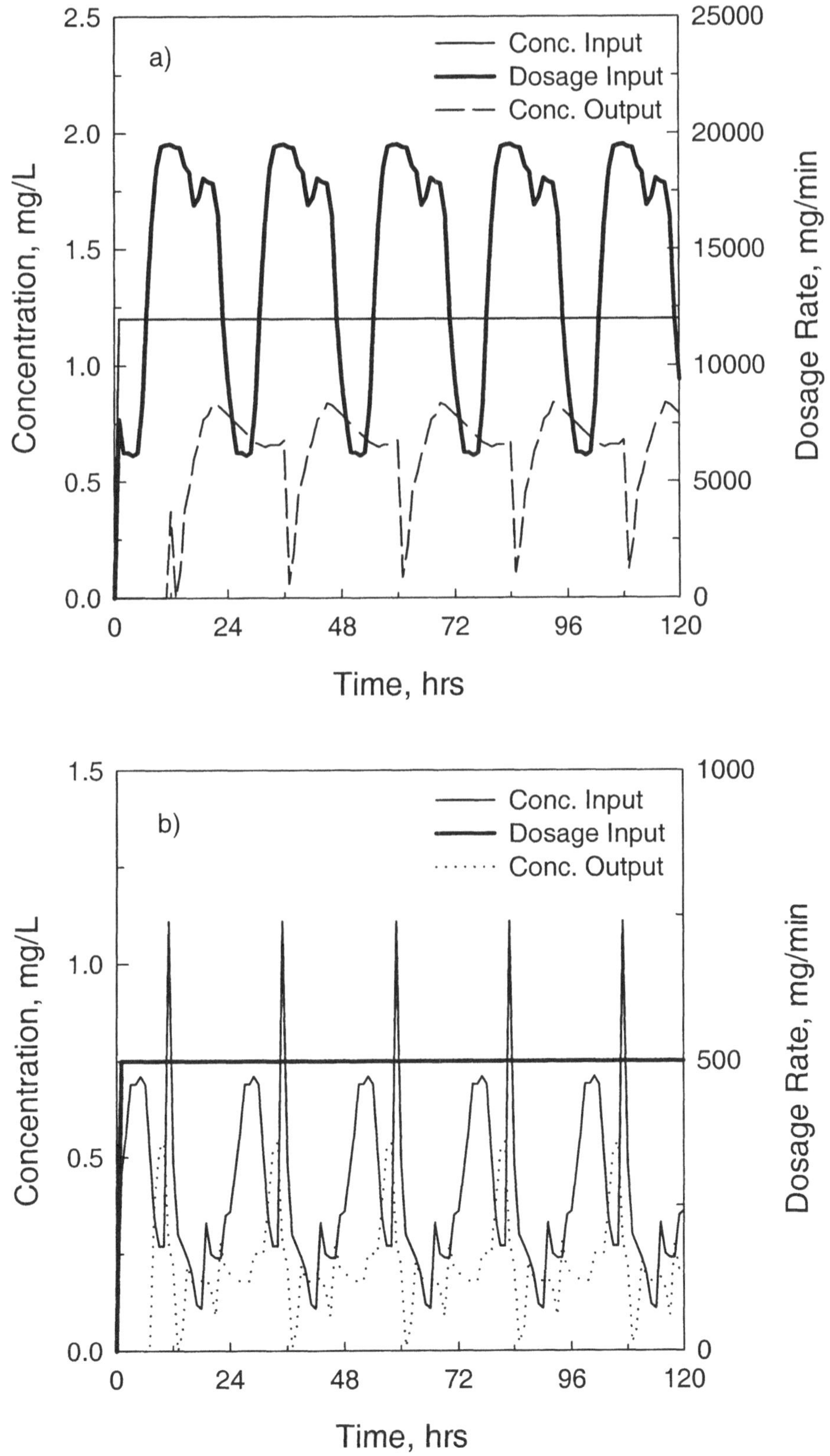

Figure 5.2: a) Injection 1, flow proportional mass dosage rate; b) Injection 2, constant mass dosage rate.

where β (M/L^3) is a constant concentration. Flow proportional dosage source types (like Injector 1) have the effect of adding the concentration β to that resulting from the mixed flows entering the node, $c_j(t)$.

Variations in input concentration can influence the output concentrations observed at locations throughout the network. Figure 5.2(a) and (b) also illustrate the output concentration at an observation location (denoted with a "+" in Figure 5.1) as a result of the dosages applied separately at Injectors 1 and 2. The output concentration due to Injector 1 (shown in Figure 5.2(a)) is generally uniform with periodic fluctuations due to drain and fill cycle of Tank B. Conversely, the output concentration due to Injector 2 (shown in Figure 5.2(b)) exhibits significant variation as periodic fluctuations of the input concentration are propagated to the output. Thus, by incorporating a flow proportional source type into the booster disinfection location and scheduling model, the advantages of reduced variation in input and output concentrations will be included in the designs.

5.1.2 Linear Superposition

Appendix D shows that the output concentrations produced under dynamic hydraulic conditions can be described using linear dynamical systems theory, assuming that disinfectant decay can be represented as a first-order reaction. Thus the output disinfectant concentrations at points within the network are a linear superposition of the responses to individual source dosage rates. Using a different approach, Zierolf et. al.(Zierolf et al., 1998) derived a recursive expression for the node concentration, and the impulse response function, by "backtracking" through the distribution system over all transport paths to find the superimposed impact of all disinfectant sources.

First, we give an intuitive explanation of linear superposition by way of a simple example. Linear superposition can be illustrated by the simple network of 4 nodes and 3 pipes shown in Figure 5.3. Water enters the system, with no disinfectant present, at nodes 1 and 2 with constant flowrates q_1 and q_2 of 1000 L/s and 2000 L/s, respectively. Water exits the system at node 4 with a flowrate of 3000 L/s $(q_1 + q_2)$. Residence times are denoted by t_{ij}, the time it takes water to travel from node i to node j, with $t_{13} = 5s$, $t_{23} = 10s$, and $t_{34} = 20s$. Disinfectant may be added at node 1 and/or node 2 with a constant mass injection rate, v_1 or v_2 respectively. Disinfectant decay is modeled by a first order reaction, where the decay constant $k = 0.05s^{-1}$.

The sample distribution system is examined under three different booster scenarios: (A) $v_1 = 2500$ mg/s, $v_2 = 0$ mg/s, (B) $v_1 = 0$ mg/s, $v_2 = 10{,}000$ mg/s, and C) $v_1 = 2500$ mg/s, $v_2 = 10{,}000$ mg/s. The concentrations at nodes 1 and 2 are calculated from their respective mass injections and flows $c_i = v_i/q_i$. Assuming first-order disinfectant decay kinetics, the concentration entering node 3 from node i is $c_i e^{-kt_{i3}}, i = 1, 2$. The concentration leaving node 3, assuming complete mixing, is the flow weighted average of its influent concentrations (i.e., $c_i = \sum_{j \in IN_i} c_j q_j / \sum_{j \in IN_i} q_j$ where IN_i is the set of upstream nodes for which flow is directed

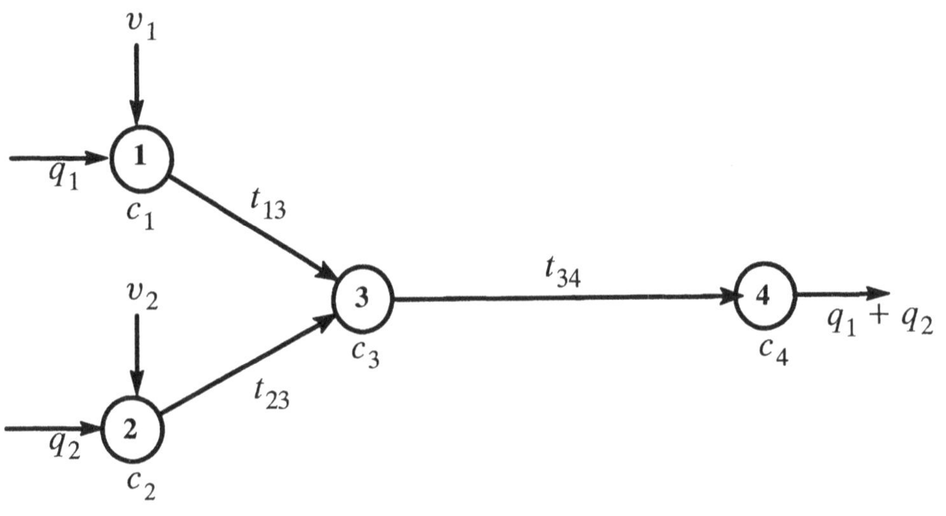

Figure 5.3: Example distribution system to illustrate linear superposition.

Table 5.1: Results for example distribution system illustrating linear superposition.

Case	v_1 (mg/s)	v_2 (mg/s)	c_1 (mg/L)	c_2 (mg/L)	c_3 (mg/L)	c_4 (mg/L)
A	2,500	0	2.5	0	0.65	0.24
B	0	10,000	0	5.0	2.02	0.74
C	2,500	10,000	2.5	5.0	2.67	0.98

into node i). The concentration at node 4 is $c_4 = c_3 e^{-kt_{34}}$. Thus the concentration at node 4 can be obtained as a linear function of the input mass rates; the proportionality constants are a_{14} and a_{24}, which define the impact on chlorine concentration at node 4 due to a unit mass rate injection at nodes 1 and 2, respectively.

The results of the three cases are shown in Table 5.1. Case A defined $c_4 = a_{14} v_1$, where $a_{14} = c_4/v_1 = 9.6 \times 10^{-5}$ ((mg/L)/(mg/s)) (given the above data). Similarly, case B defined $a_{24} = 7.4 \times 10^{-5}$ ((mg/L)/(mg/s)). Since case C applies the mass injections of cases A and B simultaneously, linear superposition states that $c_4 = 9.6 \times 10^{-5} v_1 + 7.4 \times 10^{-5} v_2$, or the summation of case A and B concentrations. Table 5.1 confirms that this is indeed the case, and linear superposition applies similarly to water quality modeling in general networks with time-varying hydraulics (see Appendix D).

Building on this simple illustration of linear superposition of booster dose effects, linear superposition can be demonstrated by simulation of concentration dynamics in a more complex network. Figure 5.4 illustrates the combined and individual concentration responses at the observation node in 5.1 (marked with a "+" symbol), due to the mass dosages applied at Injector 1 (time varying) and Injector 2 (constant). Three separate concentration time series are shown in Figure 5.4. The response to Injector 1 is the output concentration at

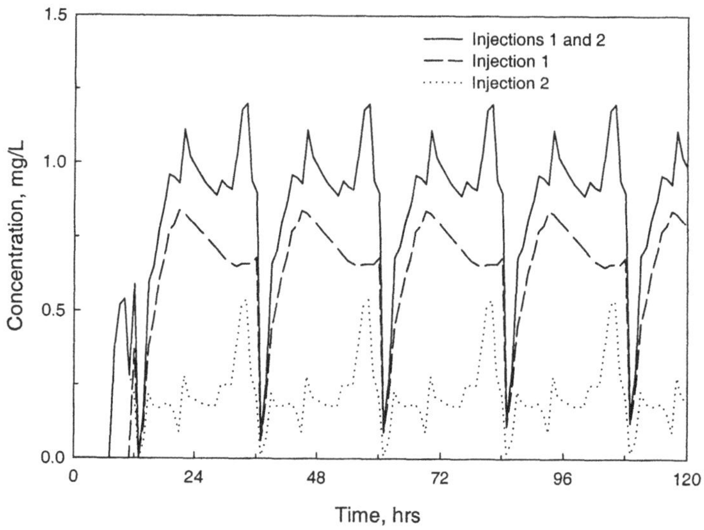

Figure 5.4: Linear superposition of output concentrations.

the observation node due to Injector 1, with Injector 2 turned off. Similarly, the response to Injector 2 is the output concentration due to Injector 2 with Injector 1 turned off, and the combined response "Injector 1 + Injector 2" is the concentration due to simultaneous dosages applied at Injectors 1 and 2. Of interest is that the two separate responses, when summed, yield precisely the combined response obtained when both injectors operate simultaneously, just as in the simple network example (see Figure 5.4).

Clearly, if the two separate responses in Figure 5.4 are available, one needn't perform another simulation to calculate their combined response; one simply adds the separate responses together. Because linear theory applies, the combined response to any multiple of dosage rates is obtained by multiplying the separate responses before adding them. For example, the combined response to doubling both dosage rates can be calculated by simply doubling the individual responses in Figure 5.4 and adding them together. This linear superposition of source dosage responses is central to the formulation of the optimal facility location model.

As the example above demonstrates, linear superposition applies to both constant and flow proportional dosage responses. Thus the linear equation relating dosages to their responses is independent of which definition for dosage rate, (5.2) or (5.3), we choose. To develop this linear equation, the disinfectant concentration at a particular network node and time is first expressed in terms of *impulse responses.* The disinfectant mass impulse at location i, during dose schedule time interval k, $I_i^k(t)$ [M], is defined,

$$I_i^k(t) = \lim_{\Delta t \to 0} \int_t^{t+\Delta t} v_i^k(\tau) d\tau \tag{5.4}$$

This mass impulse yields a concentration response, $c_j^m(I_i^k(t))$ $[M/L^3]$, at node j and time $m\delta t_m$, where m is the monitoring time step index, and δt_m is the monitoring time step. The unit impulse response coefficient, $a_{ij}^{km}(t)$ $[(M/L^3)/M]$, is the disinfectant concentration at node j and time $m\delta t_m$, resulting from a unit disinfectant mass impulse,

$$a_{ij}^{km}(t) = \frac{c_j^m(I_i^k(t))}{I_i^k(t)}. \tag{5.5}$$

Integrating these unit impulse response coefficients over all time yields a *composite* response coefficient,

$$\alpha_{ij}^{km} = \int a_{ij}^{km}(t)\upsilon_i^k(t)dt, \tag{5.6}$$

where, a_{ij}^{km} $[M/L^3]$ is the concentration response at node j and monitoring time $m\delta t_m$ to the dosage $\upsilon_i^k(t)$. The composite response coefficients can be calculated by perturbation analysis: select $\upsilon_i^k(t)$, for each individual dosage addition, as a constant or flow proportional mass dosage of arbitrary magnitude; perform a dynamic water quality simulation; and record the output concentrations.

Linear superposition allows the concentration response to multiple dosages, applied at different locations and times, to be expressed as a linear combination of individual dosage effects,

$$c_j^m(\mathbf{x}) = \sum_{i=1}^{n_b}\sum_{k=1}^{n_i} \alpha_{ij}^{km} x_i^k, \tag{5.7}$$

where n_b is the number of booster locations, and x_i^k is a dosage multiplier arising naturally from the linear response to dosage inputs. In the special case where $x_i^k = 1$ for all i, k, we are simply calculating the response to all dosages $\upsilon_i^k(t)$ acting simultaneously, corresponding to the example results in Figure 5.4. Furthermore, if $x_i^k = 2$ for all i, k, we would be calculating the response to doubling all dosages $\upsilon_i^k(t)$. Thus, the total dosage being applied in each case is equal to $\upsilon_i^k(t)x_i^k$. Eq. (5.7) is a convenient linear expression for the dependence of disinfectant concentration on multiple injector dosages, and the dosage multiplier x_i^k becomes the decision variable when optimizing dosages.

The dosage parameterization admits both constant mass rate and flow proportional dosages, as well as any other time varying dosage rate. The number of discrete booster periods, and how the dosage varies over time, depends on the level of flexibility desired when representing a disinfectant source. Boccelli et al.(Boccelli et al., 1998) used a simpler parameterization where the dosage rate was constant. Allowing for time varying dosages may seem an unnecessary complication, however, modeling and practical considerations have convinced us otherwise. As shown in the example above, a constant dosage rate can cause variations in dosage concentration at the injector that are propagated throughout the network. From a practical perspective, such variations should be avoided because they contribute to the overexposure of consumers to disinfectants, and may encourage taste and odor complaints. In the booster location model presented below, large variations in concentration at the point of injection can eliminate a location from being chosen as an injector site, if concentrations at the injector or monitoring nodes exceed constraint upper bounds. Flow proportional dosages help to alleviate these problems associated with constant mass dosage rates. We note that while a constant mass dosage rate u_i^k could represent a time varying dosage via a time discretization that matches that of the hydraulic solution, the price would be a significant increase in the number of multipliers x_i^k, and thus the computational work required to set up and solve the location model.

5.1.3 Optimization Model Formulation

The booster facility location model selects a set of injector locations to minimize the average disinfectant mass dosage rate applied to the network. Constraints ensure that disinfectant concentration is maintained within lower and upper bounds at selected monitoring nodes. As is typical of facility location models, binary variables represent the decision to build, or not, at a particular location. The number of booster stations and the disinfectant mass are related to the cost of constructing and operating a booster disinfection system. Also, minimization of disinfectant mass applied, while maintaining minimum residual concentrations, implies minimization of the disinfectant mass lost to bulk reactions with natural organic matter and to wall reactions with pipe materials. Since chlorine consumption has been related to DBP production (see chapter 4, also (Clark,)), minimization of disinfectant mass dosage rate may also be a surrogate for minimization of regulated DBPs.

Mathematically, the optimal facility location model takes the form,

$$\min_{\mathbf{x},\delta} \sum_{i=1}^{n_b} \left(\frac{1}{\Delta T_i} \sum_{k=1}^{n_i} M_i^k x_i^k\right), \tag{5.8}$$

where $\Delta T_i = \sum_{k=1}^{n_i} \delta t_i^k$ [T] is the length of the ith schedule cycle, and M_i^k [M] is the disinfectant mass associated with the periodic dosage rate $v_i^k(t)$,

$$M_i^k = \int_0^{\delta t_i^k} v_i^k(t)dt. \tag{5.9}$$

The term in parentheses (5.8) is the average mass rate for the dosage at location i. This minimization is subject to constraints on minimum (l_j) and maximum (u_j) disinfectant concentrations at n_m monitoring nodes and n_α monitoring time steps (see (5.7)),

$$l_j \leq \sum_{i=1}^{n_b} \sum_{k=1}^{n_i} \alpha_{ij}^{km} x_i^k \leq u_j, \quad j = 1, \ldots, n_m, \; m = M, \ldots, M + n_\alpha - 1, \tag{5.10}$$

where $M\delta t_m$ is the start of the monitoring time window (discussed below). One binary variable δ_i is introduced for each potential booster location, indicating whether ($\delta_i = 1$), or not ($\delta_i = 0$), a booster station is to be constructed at i. Constraints ensure that the mass rate at location i is zero if a booster station is not constructed,

$$x_i^k \leq X_i^k \delta_i, \quad i = 1, \ldots, n_b, \; k = 1, \ldots, n_i, \tag{5.11}$$

where X_i^k is a constant equal to the maximum feasible value of x_i^k (discussed below). The maximum number of booster stations is n_b^{max},

$$\sum_{i=1}^{n_b} \delta_i \leq n_b^{max}, \tag{5.12}$$

and the δ_i are restricted to binary and the x_i^k to non-negative values,

$$\delta_i = \{0, 1\}, \quad x_i^k \geq 0, \quad i = 1, \ldots, n_b, \; k = 1, \ldots, n_i. \tag{5.13}$$

Minimization of (5.8) subject to constraints (5.10-5.13) constitutes the MILP booster facility location model, which includes continuous decision variables x_i^k, $i = 1, \ldots, n_b$, $k = 1, \ldots, n_i$, and binary decision variables δ_i, $i = 1, \ldots, n_b$.

5.1.4 Monitoring Time Window

The booster facility location model considers capital and operating decisions for a dynamic system with uncertain future inputs over an extended useful life; as such, it represents the typical planning process. As with most planning exercises, assumptions about the future are needed to relate the decisions made today to meaningful measures of system performance in the future. In the present case, these assumptions relate to the dynamics of the disinfectant concentration response. Below we give a summary of the approach, which is discussed in greater detail by Boccelli et al.(Boccelli et al., 1998).

The disinfectant concentration dynamics at each monitoring node – the superposition of multiple booster doses - is assumed to be periodic for $t \geq M\delta t_m$ (aperiodicity for $t < M\delta t_m$ is due to initial condition effects). Further, the length of one concentration cycle is assumed no greater than $n_\alpha \delta t_m$. If these hold, then the monitoring constraints (5.10) ensure that disinfectant concentrations are within lower and upper bounds for all time $t \geq M\delta t_m$.

If the concentration dynamics are to be periodic, then other system characteristics must be chosen carefully. At least, one requires a periodic dosage rate like that resulting from the parameterization described above. The system hydraulic dynamics, and thus the system demands that drive them, also must be periodic. Apparently there is no general proof of hydraulic periodicity given demand periodicity, although it appears to be an implicit assumption made in many practical studies, such as where a particular diurnal variation (e.g., maximum day) is selected for long term capital planning. In the absence of such proof, a practical method is to perform a simulation check for periodicity of hydraulic and concentration dynamics (the approach used in the applications in chapter 6).

5.1.5 Solution Procedure

A program – Booster Disinfection Design and Analysis (BDDA) – was written that combines the problem setup and solution phases of the booster facility scheduling and location model;

a User's Guide for the BDDA is provided in Appendix A. The problem setup phase uses the EPANET Programmers Toolkit(Rossman, 1999) to compute the EPS network hydraulics and the composite response coefficients. Provided that the assumptions related to periodicity and linear superposition are satisfied, the α_{ij}^{km} coefficients can be computed by perturbation using the algorithm used by Boccelli et al.(Boccelli et al., 1998) and summarized here.

1. Select the desired source type, dosage discretization, and a suitable value for $v_i^k(t)$ at injector i.

2. While keeping all other booster injections set to zero, simulate the concentration response due to dosage $v_i^k(t)$.

3. Run a long-term water quality simulation over many cycles of the resulting periodic concentration dynamics until the mean concentration becomes stationary.

4. Record $\alpha_{ij}^{km} = c_j^m$ for all monitoring locations $j = 1, \ldots, n_m$ at each monitoring time $m = M, \ldots, M + n_\alpha - 1$.

5. Repeat the process for all booster periods $k = 1, \ldots, n_i$ and all injector locations $i = 1, \ldots, n_b$.

Thus, α_{ij}^{km} is simply the simulated concentration response, c_j^m, to the dosage $v_i^k(t)$, regardless of whether is is constant or flow proportional. It is not necessary for the designer to manually specify the dosage rate variations $v_i^k(t)$ in step 1 of the algorithm; such a task would be tedious. Rather, the designer specifies what type of disinfectant source model is desired (constant mass rate or flow proportional mass rate), and water quality simulation algorithms take care of the rest. The Epanet Programmer's Toolkit(Rossman, 1999) includes the ability to model constant sources by specifying a constant mass dosage rate, or flow proportional sources by specifying the concentration addition at the injector (β, see (5.3)). In either case the user specifies n_i and δt_i^k, selects a suitable magnitude for the dosage (γ or β), and the Epanet source algorithm calculates $v_i^k(t)$ automatically during the simulation. As such, selecting between dosage types is trivial from the perspective of the designer.

The objective function coefficients M_i^k are computed as part of the algorithm by accumulating the disinfectant mass added at an injector over a schedule cycle according to (5.9). The above procedure requires one water quality simulation for each decision variable x_i^k, so problem setup time scales linearly with the number of decision variables. This time can be significant for practical problems, due to the time required for simulation of network water quality dynamics. The BDDA code, however, was designed so that the response coefficients can be stored and reused, for solution of problems that are similar except for constraint right hand sides (e.g., changing the number of stations allowed to operate, or the bounds on monitored residual concentrations).

The value of each constant X_i^k (5.11) is calculated using the composite response coefficient by considering limits on x_i^k imposed by the lower and upper bounds on monitored concentrations. The X_i^k must not restrict the optimal value of x_i^k, yet it should not be needlessly large; a carefully selected value will force the relaxed (LP) optimal solutions (part of the branch–and–bound solution procedure) to reflect more of their true cost, and thus may lead to a more efficient branch and bound search. Let $\overline{x_i^k} = \min(u_j/\alpha_{ij}^{km}), \forall jm$, the maximum feasible value of x_i^k based on upper concentration bounds (assuming, conservatively, that x_i^k is the only non-zero dose multiplier). Similarly, let $\underline{x_i^k} = \max(l_j/\alpha_{ij}^{km}), \forall jm$, the minimum value of x_i^k such that all lower concentration bounds are satisfied. The value of X_i^k is safely taken as the minimum of these two values, $X_i^k = min(x_i^k, x_i^k)$, since minimization of (5.8) implies that $x_i^k \leq \underline{x_i^k}$ for any optimal solution.

The time associated with network simulations during problem setup is proportional to $n_b n_i$. The problem solution phase is performed using CPLEX (see Appendix E), a commercial branch and bound procedure using the simplex algorithm for integer relaxed subproblems(Efroymson and Ray, 1968; Nemhauser and Wolsey, 1988). The program interfaces with CPLEX through its callable library (for automatic solution after problem setup), or it writes a standard MPS format input file compatible with other commercial or public domain solvers. Problem solution time using the branch–and–bound method depends on the number of binary decision variables, the size of the LP sub problems, optimality criteria, and various other algorithmic options and tolerances. Practical application of the model to networks with thousands of nodes may require hundreds of potential booster locations, thousands of monitoring locations, and monitoring times that capture daily variation in disinfectant residual – leading to formulations with tens of thousands of constraints, and hundreds of continuous and binary decision variables. Thus, while experience indicates the model can be solved for practical networks, its application should not be considered routine at this time.

5.1.6 Summary

A method is developed which allows the optimal location and scheduling of booster injection stations in drinking water distribution networks. The model formulation is related to the general fixed-charge facility location problem and expresses the decision to build or not build at a potential location using binary variables. Time varying dose rates are incorporated to allow efficient scheduling of doses if booster flow rate variation is significant. Because the problem is expressed in terms of both continuous scheduling and binary location variables, the formulation is a mixed integer linear program, and a branch and bound algorithm is used for solution.

5.2 Maximum Set Covering Location Model

Here we show how an alternative booster station location problem can be formulated as a classical integer linear programming (ILP) model called the maximum set covering model (MSC). Such a formulation brings with it significant computational advantages for medium to large networks; it may be possible to consider all nodes as potential booster locations, while maintaining a computationally tractable model. Also, efficient heuristic solution methods are available that do not require commercial MILP solvers. These computational advantages require that we give up optimization of dose schedules, and instead treat all booster stations as if they are operated in a typical fashion. So, unlike the MILP formulation above, the MSC formulation does not build upon the optimal scheduling (LP) model(Boccelli et al., 1998) directly, and thus does not combine the two functions of optimal scheduling and optimal location. We suggest the MSC model can be used as a screening model in concert with either the optimal scheduling or optimal scheduling/location model formulations presented earlier.

5.2.1 Optimization Model Formulation

The MSC formulation draws upon results from facility location models, and in particular the maximal covering location model(Church and Revelle, 1974), which has also provided inspiration for models that locate wells in groundwater monitoring networks(Meyer and E. D. Brill, 1988). Consider a number of discrete potential booster station locations (n_b), and discrete monitoring locations (n_m) where disinfectant residual is to be controlled. As with the above MILP formulation, for design purposes the network flow rates and disinfectant concentrations are assumed to be time dependent and periodic, and we sample the disinfectant concentrations over a set of discrete time intervals, $k = 1, 2, \ldots, K$, corresponding to one period of the dynamics. Further, define a set of potential booster locations, N_j^k, such that booster location i is included in N_j^k if and only if disinfectant dosing at i elicits a "significant response" at sensor location j and time interval k. Thus N_j^k is the set of all potential booster locations that are assumed to have an effect on residual at sensor location j and time k. We discuss the definition of the sets N_j^k below, as they are central to the formulation of the booster location model as a maximal set covering model.

Given the above definitions, we seek to minimize the number of booster stations (a surrogate objective for minimizing the cost of booster station installation and operation),

$$\min_{\delta} \sum_{i=1}^{n_b} \delta_i, \tag{5.14}$$

where δ_i is a binary variable equal to 1 if a booster station is to be constructed at location i, and zero otherwise. In minimizing the number of booster stations, we require that each monitoring location and time include a significant response from *at least one* selected booster location, else the residual concentration at that location and time would be difficult to control using the booster dosages. Mathematically, these constraints are,

$$\sum_{i \in N_j^k} \delta_i \geq 1, \; j = 1, \ldots, n_m, \; k = 1, \ldots, K. \tag{5.15}$$

Finally, the δ must be binary,

$$\delta_i = \{0, 1\}, \ i = 1, \ldots, n_b. \tag{5.16}$$

Solution of (5.14) subject to (5.15) and (5.16) will yield the minimum number of booster locations that provide spatial and temporal "coverage" for the network, in that the residual concentration at each monitoring location and time is influenced by at least one selected booster location.

5.2.2 Dose Response and Significance Level

In the above development we introduced the notion of whether, or not, dosing at a booster location elicits a significant response at a monitoring location and time. This notion of concentration "coverage" is addressed here, along with definition of the sets N_j^k.

Define the concentration response coefficients, α_{ij}^k ($[M/L^3]/[M/L^3]$), as the disinfectant concentration observed at monitoring node j and time k, due to maintaining a *unit set-point concentration* at booster location i for all time $m < k$. (These response coefficients are similar to those defined for the above MILP formulation, and are calculated similarly, except they represent the response to a constant booster output concentration rather than a periodic dosage rate.) Thus the concentration at location j and time k due to an historic set-point concentration ν_i at booster location i is simply $c_j^k = \nu_i \alpha_{ij}^k$. As above, this development assumes that disinfectant decay in the distribution system behaves as a first-order reaction (a common assumption), in which case linear superposition implies linearity of dose response.

One measure of whether, or not, a dose at booster location i exerts a sufficient response at monitoring node j and time k is simply whether, or not, $c_j^k = \nu_i \alpha_{ij}^k > \hat{c}$, where $\hat{c}$ (M/L^3) is an assumed minimum concentration. For example, we might assume that an individual booster station contribution less than 0.05 mg/l is "insignificant" ($\hat{c} = 0.05$ mg/l). In the absence of *a priori* information to suggest otherwise, we assume the same set point concentration, $\hat{\nu}$, for all potential booster locations, so that $c_j^k = \hat{\nu}\alpha_{ij}^k$. Our definition of "significant response" is then $\hat{\nu}\alpha_{ij}^k > \hat{c}$, or more simply $\alpha_{ij}^k > \eta = \hat{c}/\hat{\nu}$. The parameter η is a *coverage parameter*, interpreted as the minimum allowable monitoring node concentration that results from a unit set-point concentration at a booster station. The larger the value of η, the larger the response at the sensor must be in order to be considered "covered" by that booster station. Mathematically, the notion of coverage is reduced to whether or not the response coefficients exceed the threshold η.

Using the above definition, we define the significance sets $N_j^k(\eta)$ as the set of all booster locations having a dose response at monitoring location j, and time k, that exceeds the coverage parameter η,

$$N_j^k = \{i | \alpha_{ij}^k > \eta\} \tag{5.17}$$

The residual coverage parameter will be a key determinant of solutions to the MSC booster location model. As a thought exercise, one may consider a network where every node (and thus every monitoring location) is a potential booster location. The optimal MSC solutions for the extreme values of $\eta = \epsilon$, where ϵ is an arbitrarily small strictly positive constant, and $\eta = 1$ are known immediately. When $\eta = \epsilon$ a minimal set of booster locations is comprised of those associated with network water sources (this is the conventional case where disinfectant is added only at treatment nodes). When $\eta = 1$ a minimal set of locations is every node in the network, which is the only way to have a perfectly uniform concentration at each monitoring node. Thus η will evidently have a marked effect on the number of booster stations required to achieve "coverage" of the monitoring locations and times, and a plot of the optimal number of booster stations (from solution of the MSC problem) versus the residual coverage parameter indicates the optimal tradeoff between variability of disinfectant residual and a measure of residual maintenance cost. Such a tradeoff is presented for one example application in chapter 6.

5.2.3 Solution Procedure

The maximum set covering model formulation is in the class of integer linear programming (ILP) or mixed integer linear programming (MILP) optimization models. The model can be solved rigorously via a branch and bound solution procedure. Software that implements this solution procedure has been written and is included with the Booster Disinfection Design and Analysis (BDDA) software (see Appendix A) that interfaces with network simulation software (the EPANET programmers toolkit (Rossman, 1999)) and standard linear programming algorithms. This software is applicable to general water distribution systems; using a network model it automatically generates all response coefficients needed to construct the residual coverage sets, which are then stored in a binary file for use during subsequent model formulation. The model can then be automatically solved via a branch-and-bound procedure if desired. Maximum set covering models often are, however, "integer friendly" in that the linear programming relaxation (the form of the optimization model where all binary variables are allowed to vary continuously between 0 and 1) is naturally integer. This has been the experience for all models solved to date by the project team, and as a result the rigorous branch-and-bound solution approach is surprisingly efficient, even for relatively large problems (large numbers of integer variables).

Maximum set covering models also lend themselves to approximate solution by a simple greedy heuristic method that does not require formal optimization methods. The greedy algorithm proceeds as follows.

1. Initialize the set of booster locations that have been selected to the empty set, $S_b = \emptyset$, and the set of booster locations that have not been selected to all potential locations, $N_b = \{i|i = 1, \ldots, n_b\}$. Initialize the set of sensor controllability constraints (5.15) that are satisfied to the empty set, $S_c = \emptyset$, and the set of controllability constraints that have not been satisfied to all sensor locations and times, $U_c = \{(j,k)|j = 1, \ldots, n_m,\ k = 1, \ldots, K\}$.

2. Rank all booster locations in U_b by the total number of sensor controllability constraints $(i, j) \in U_c$ that it satisfies.

3. Remove the top ranked booster location (the one that satisfies the largest number of controllability contstraints in U_c) from U_b and add it to S_b. Remove the corresponding controllability constraints from U_c and add them to S_c.

4. If $U_c = \emptyset$ or $U_b = \emptyset$ then `STOP`. Else, go to step 2.

As with most greedy algorithms, this one is in general non–optimal. On the other hand it is easy to implement, does not need special purpose software for execution, and has yielded excellent solutions for all cases studied by the project team. The heuristic solution procedure is also included with the BDDA software (selectable via the `HEURISTIC` keyword in the `[OPTIONS]` section of the BDDA input file), and thus allows for booster location solutions without commercial linear programming software.

Chapter 6

Example Booster System Designs

This chapter presents applications of the optimal scheduling and location models to three different distribution networks. The examples cover: 1) design of booster dose schedules given that location are known, 2) simultaneous selection of booster dose schedules and booster locations, and 3) heuristic screening of potential booster locations. In addition the results illustrate some of the potential advantages of booster disinfection, and modeling issues to be aware of when using network models for design and analysis of booster systems. All results presented here were obtained using the Booster Disinfection Design and Analysis (BDDA) software (see Appendix A), which relies on the Epanet Toolkit(Rossman, 1999) to perform network hydraulic and water quality analyses.

6.1 SCCRWA Optimal Scheduling Analysis

The booster scheduling results presented here serve as an illustration of the optimal design method presented in chapter 5, and illustrate some of the potential advantages of a booster disinfection residual maintenance strategy. The example presented is for a well-studied and relatively simple water distribution netowork model derived from the Cherry Hill - Brushy Plains portion of the South Central Connecticut Regional Water Authority (SCCRWA) distribution network. This particular network has been used in many water quality modeling studies(Biswas et al., 1993; Clark et al., 1993; Rossman et al., 1994), and will be a useful first example because the network hydraulics are relatively simple to describe and understand (compared to the networks used for the following two applications). The booster station locations are assumed, and the design problem is to determine their best mode of operation – defined as that which minimizes the total disinfectant mass applied while meeting residual maintenance constraints. Booster location issues are investigated, but only via a sensitivity analysis where the locations are changed to illustrate location–dependent effects. The effect of system hydraulic dynamics on booster system design is also illustrate via a sensitivity analysis, assuming different idealized modes of hydraulic operation (or, assumptions about operation). The optimization results were obtained using the BDDA software and the `FORMULATION MINMASS` directive in the `[OPTIONS]` section of the design input file (see Appendix A). The SCCRWA network also serves as the booster station design example provided in the BDDA User's Guide (Appendix A, section A.5).

The network is shown in Figure 6.1, and includes 34 consumer nodes, 6 booster nodes, one source node representing a pump station, one storage tank and 47 links representing 7 miles of pipe. Consumer nodes (nodes 2-25, 27-36) represent water demand locations while nodes A–F represent booster disinfection station locations (some or all of these will be used, depending on the analysis). The cylindrical tank, node 26, was modeled as a continuous flow stirred tank reactor and has a diameter of 50 ft with minimum and maximum allowable water levels of 50 ft and 70 ft (above the bottom of the tank), respectively.

6.1.1 Network Hydraulic Behavior

The network hydraulic behavior is shown in Figure 6.2, where flows into the system are negative and flows out of the system are positive. The hydraulics were periodic on a 24 hour cycle; as a result, the total demand equals the total supply over the same period. The flow into and out of the tank was equal to the difference between the source inflow and consumer demands at each time step; a negative flowrate indicates the tank was draining and a positive flowrate indicates the tank was filling. Total hourly demands vary between nearly 800 and 100 gpm. The source and tank flows occurring over the 24 hour cycle divide the hydraulic dynamics of the network into four distinct periods. During periods 1 (0 - 6 hrs.) and 3 (12 - 18 hrs.) the network hydraulics were controlled mainly by the source, with flows varying between 100 and 690 gpm, transporting water up the main trunk and filling the tank. During periods 2 (6 - 12 hrs.) and 4 (18 - 24 hrs.) the pump was off and flows are controlled by the tank, causing reversal of flow in the main trunk between nodes 2 and 25.

6.1.2 Experimental Design

Booster locations A - F (see Figure 6.1) were chosen based on trial simulation results obtained for the modelled SCCRWA network. Booster node A simulates boosting at the source of the network. Node B is located off the main trunk to serve the stem from node 7 to node 10. Node C is located along main trunk (near node 7), serving the stems and loops above and below its location depending on flow direction. Booster nodes D and E serve the upper most reaches of the network, while boosting at the stem of the tank is provided by node F. Locations B and C, and D and E occupy essentially identical geographic locations. Nodes B and C were included in anticipation of flow reversal patterns that occur along the main trunk line. For example, during periods 1 and 3, (when the pump is on) booster C has no effect (during the same period) on the nodes below it leaving boosters A and B to satisfy residual constraints. During periods 2 and 4, however, booster station C can contribute

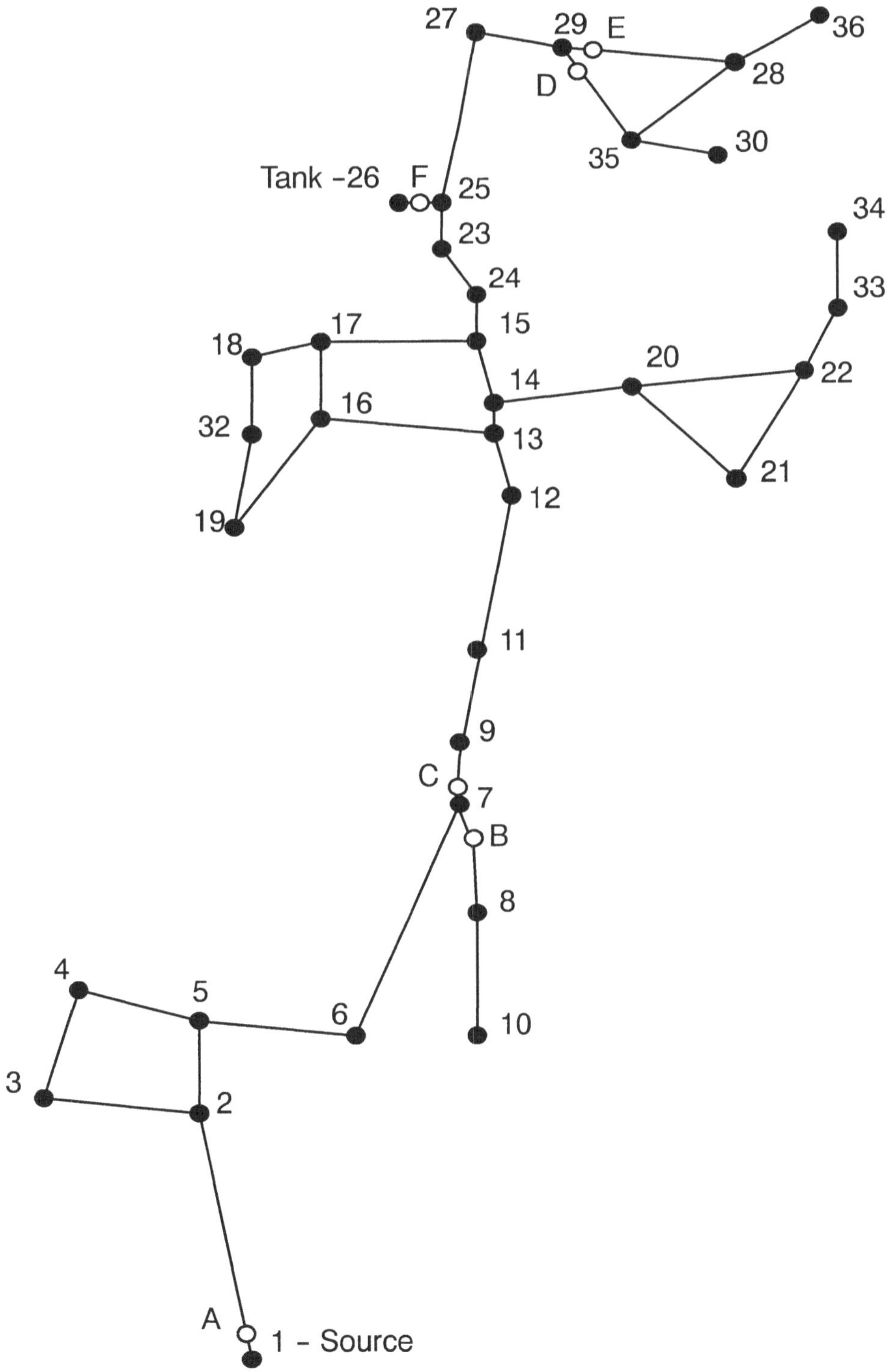

Figure 6.1: SCCRWA distribution system schematic.

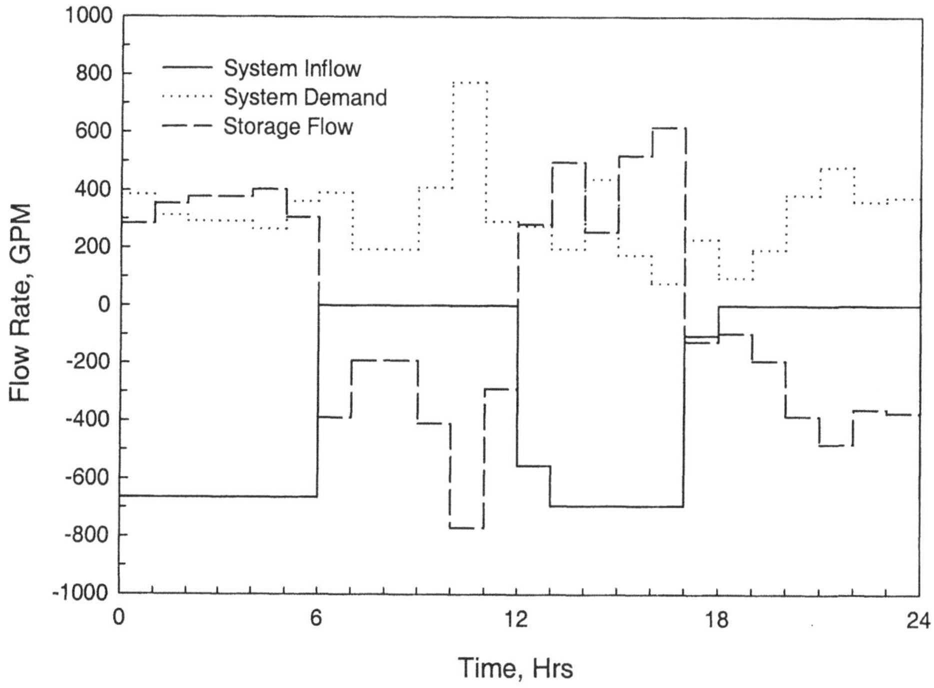

Figure 6.2: SCCRWA network hydraulic behavior for cases I-III

to disinfectant residuals in both lower stems during the same period, and is the closest disinfectant source to the loop near the pump (nodes 2-5).

Disinfectant concentrations were monitored at all consumer nodes to ensure that residual constraints were satisfied throughout the distribution system. The values of l_j and u_j were chosen as 0.2 mg/L and 4.0 mg/L, respectively, to reflect current best practices and anticipated SDWA regulations. The booster schedule cycle time (T_s) was chosen to coincide with the hydraulic cycle time of 24 hours (T_h) ($\eta = \mu = 1$). The booster stations were modeled as `MASS` sources, with a constant mass injection rate over each of four 6 hour periods, set to coincide with the four distinct periods of system hydraulics; thus the operation of each booster station is described by four separate constant mass injection intervals, each six hours long, that repeat every 24 hours. The disinfectant decay rate constant (k_b) used for the simulations was 0.5 d^{-1}, and only considered the bulk decay, as wall decay was assumed negligible. The BDDA design input file for this example is listed in Table A.4.

Five simulation cases, shown in Table 6.1, were investigated to illustrate how the optimal schedule is influenced by the number and location of booster stations, and the system hydraulics. Cases I – III examine the influence of the number and location of booster nodes on the optimal schedule results under the dynamic hydraulic conditions described above. Case I considers disinfectant addition only at booster A, a conventional disinfection scenario. For case II, four more booster nodes were added at locations B through E. For case III, an additional booster node was added at the tank, node F.

Table 6.1: SCCRWA simulation cases.

Case	Booster Locations	Demands	Pump Operation
I	A	Dynamic	Dynamic
II	A,B,C,D,E	Dynamic	Dynamic
III	A,B,C,D,E,F	Dynamic	Dynamic
IV	A,B,C,D,E	Dynamic	Steady
V	A,B,C,D,E	Steady	Steady

Cases IV and V examine the impact of system hydraulics on the optimal schedule for the booster nodes assumed for case II. Case IV assumed hydraulic conditions associated with a steady pump flow rate as shown in Figure 6.3. The steady pump flow rate was calculated by averaging the dynamic pump flow rates over a 24 hour period, resulting in a constant flow rate of 310 gpm. Steady-state hydraulics were assumed for case V, corresponding to the average steady-state demand and pump flows. In this case, the tank plays no role in determining the optimal schedule, as there is no flow into or out of the tank.

6.1.3 Results and Discussion

The case I results, shown in Table 6.2, show large injection rates at node A during periods 1 and 3. These large injection rates were required so that enough disinfectant enters the distribution system, especially the tank, to ensure an adequate residual is maintained when the pump is off and the tank feeds the system. The absence of injections during periods 2 and 4 was due to the pump being off during those times. The total mass applied for case I was 2824 kg/d. This case was equivalent to conventional treatment since disinfectant was added only at the source, and thus will be used as a basis of comparison for cases II and III.

The case II results (Table 6.2) show large injection rates at nodes A and C during periods 1 and 3, in which the pump was operating, with much smaller injections at nodes B and E. Node C also was operating with a large injection rate during period 2, when the tank dominates the flow. No injections occurred during period 4 indicating disinfectant concentration requirements were being satisfied by the disinfectant concentration present in the storage tank and resident in the system. The addition of the four booster nodes (B-E) in case II reduced the total mass applied from case I to 2305 g/d, an 18.4% reduction, although significant booster injections occurred only at booster locations A and C.

The case III results (Table 6.2) show large injection rates occurring at nodes A and C during periods 1 and 3, in which the pump is operating, with much smaller injections at nodes B, D, and E. The large injection rate at node C during period 2, for the case II results, has been replaced by injections at node F during periods 2 and 4 corresponding to the drain cycles of the tank. This shift in the location and time of boosting allowed a

Table 6.2: SCCRWA optimal scheduling results.

Case	Total Mass Rate (g/d)	Booster Period†	Booster Loc. / Inj. Rate (mg/min)					
			A	B	C	D	E	F
I	2824	1	5678.9	-	-	-	-	-
		2	0.0	-	-	-	-	-
		3	2166.0	-	-	-	-	-
		4	0.0	-	-	-	-	-
II	2305	1	587.7	2.7	3243.5	0.0	0.0	-
		2	0.0	0.0	335.3	0.0	0.0	-
		3	1421.4	0.0	811.9	0.0	0.0	-
		4	0.0	0.0	0.0	0.0	0.0	-
III	1120	1	587.7	4.4	354.0	0.0	0.1	0.0
		2	0.0	0.0	0.0	0.3	0.4	671.1
		3	634.7	4.5	459.3	0.0	0.2	15.0
		4	0.0	0.0	0.0	0.2	0.2	377.9
IV	3475	1	250.5	5.2	3418.5	0.0	0.0	-
		2	499.3	5.0	1514.7	0.0	0.2	-
		3	274.8	2.7	3237.4	0.0	0.1	-
		4	250.9	5.7	187.8	0.05	0.3	-
V‡	697.1	1	249.8	4.5	229.8	0.03	0.03	-

†Booster period: 1 = hours 0–6, 2 = hours 7–12, 3 = hours 13–18, 4 = hours 19-24

‡Injection rates for booster periods 2, 3, and 4 are identical to booster period 1.

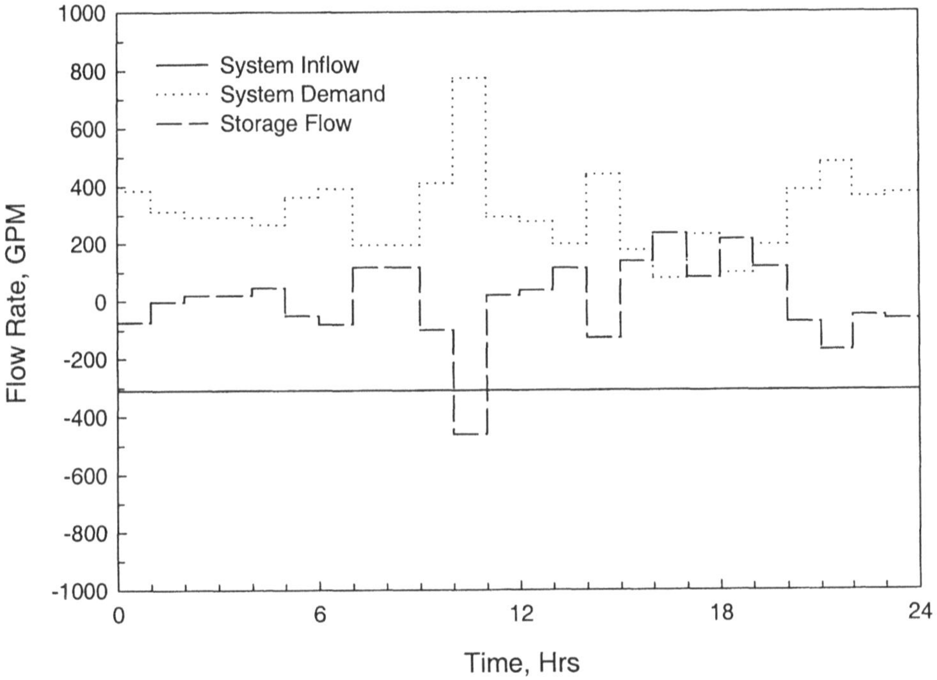

Figure 6.3: SCCRWA network hydraulic behavior for case IV

significant reduction in disinfectant required with a total mass applied for case III of 1120 g/d, a 60.3% reduction over case I.

The case IV results (Table 6.2), compared with those for case II, show large injection rates occurring at nodes A and C during all hydraulic periods, because the pump is operating at a constant flow rate. Since the demands are identical to those for case II, but the pump flow rate was constant, the tank drain/fill cycle was more irregular than for case II (see Figure 6.2). The total mass rate required to satisfy residual constraints was equal to 3475 g/d, a 50.8% increase over case II.

The case V results (Table 6.2) can be described by one mass injection rate per location, since the hydraulics were constant over all four booster periods; thus the optimal rates are identical for each period. Large injections occurred at nodes A and C, similar to the case IV results. The steady state hydraulics result in a substantial decrease in the disinfectant required with a total mass applied equal to 697.1 g/d, a 69.8% reduction over case II, and significantly lower than any other case examined.

The steady-state hydraulic results deserve emphasis. Essentially, the steady-state assumption eliminates the large volume of water stored in the tank, as well as the disinfectant dilution and decay effects associated with it. Under this hydraulic simplification, the impact of a booster injection was exaggerated, relative to its impact under more realistic dynamic hydraulics. The steady-state hydraulic analysis was clearly an over-simplification for this particular problem, and should not be used as a basis for design of booster injection schedules

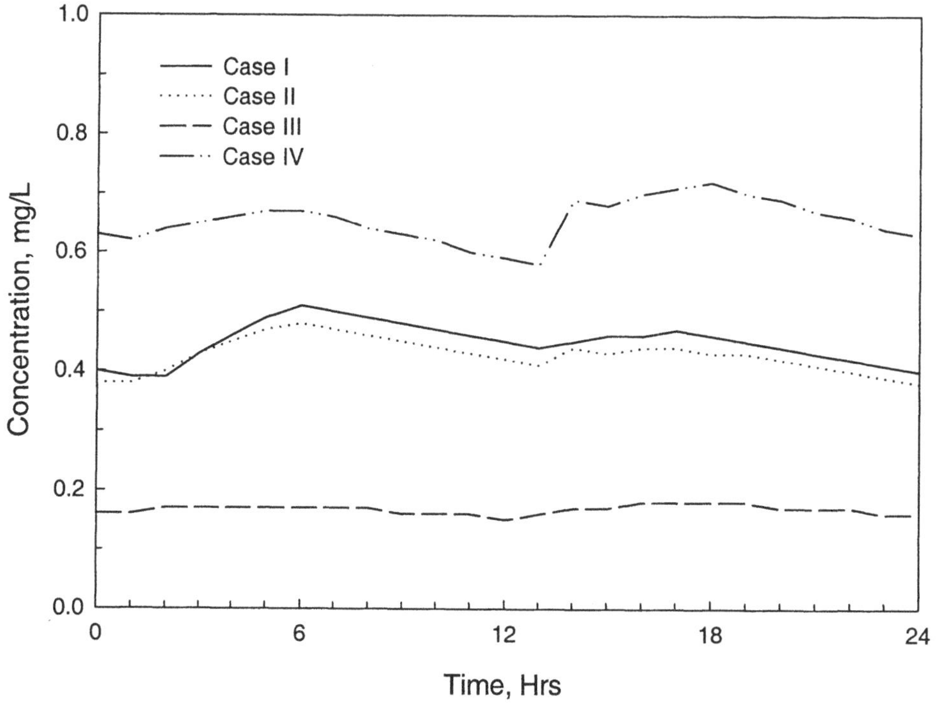

Figure 6.4: SCCRWA tank concentration versus time for cases I-IV.

(or for any water quality analysis, for that matter). If the case V results were implemented under dynamic hydraulics, the mass applied to the system would fail to maintain adequate disinfectant residuals.

The ability of booster disinfection to reduce the total amount of disinfectant required to satisfy monitoring constraints depends, to some extent, on the disinfectant concentration dynamics produced in the tank. The time varying disinfectant concentrations present in the tank for cases I - IV are shown in Figure 6.4. Cases I and II had similar concentration profiles; disinfectant concentrations increase in periods 1 and 3, when the pump was operating, and decrease in periods 2 and 4, when the tank was draining. This is due to the similarity of the optimal schedules produced by the two cases, in particular, the large injections occurring at nodes A or C in the first and second boosting periods. These large injections were stored in the tank and used to maintain disinfectant residuals during the tank's drain cycle.

The tank concentration profile produced by case III (Figure 6.4) was less than 0.20 mg/L. It should be noted, however, that constraints were not written for disinfectant residuals present in the tank (although these could be included using the BDDA software), and a lower tank residual increases the potential for bacterial growth and nitrification (with chloramines). The booster injections at node F, during the tank's drain cycle, removed the need to use the tank as a disinfectant source; thus the total amount of disinfectant required to satisfy the monitoring constraints was reduced by minimizing the tank concentration, and thus the loss of disinfectant in the tank. Disinfectant decay in the tank can be large, due to the tank's large residence time, and the dilution factor associated with such a large volume of stored water.

Summarizing, the results of cases I - III suggest that booster disinfection can reduce the amount of disinfectant needed to maintain a residual, the location of these boosters is significant in determining the amount of mass reduction, and storage tanks effects on residence time can be important to consider when determining good booster locations and operation schedules.

Case IV (Figure 6.4) produced a tank concentration profile greater than for cases I - III. This increase in concentration was required to satisfy monitoring constraints within the distribution system, thus increasing the mass of disinfectant applied to adequately maintain these higher tank concentrations. As mentioned earlier, the drain/fill cycle of the tank was made more irregular due to the steady-state pump flow rate. This irregularity decreased the volume of water entering the tank, requiring that even more disinfectant mass be applied in order to achieve the necessary tank concentration profile.

A measure of network wide concentration variability was used to determine if booster disinfection could produce a more uniform residual with less total disinfectant applied. Figure 6.5 through 6.7 show the average and standard deviation of observed concentrations at all monitoring nodes in the system taken each hour for cases I through III respectively. While the addition of booster stations B - E did decrease the average concentration from case I to case II, the standard deviations increased, indicating large spatial variations in concentration exist for both cases I and II. Addition of booster station F in case III (Figure 6.7) produced a pronounced drop in both the average and standard deviation of concentration, implying a more uniform spatial and temporal distribution of disinfectant concentrations in the network.

The disinfectant concentration dynamics were a complex function of the dynamic hydraulics and the optimal schedule of booster injections. Figure 6.8 shows the concentration over time at node 36, for the case II optimal schedule. The Figure was developed such that the contribution of each booster mass injection has been displayed cumulatively to give the total concentration present at the node; consequently, the contribution of an individual booster mass injection is the area between its line and the next line on the figure, and the top line is the cumulative concentration due to all individual injections. These results reveal that disinfectant concentrations result from a complex combination of individual injection impacts from different booster locations and periods, and that different disinfectant mass injections contributed significantly at different times. Nodes A, C, and F all make significant contributions to the total concentration present.

Further examination of Figure 6.8 reveals that the arrival of contributing booster injections were delayed in time. Considering node F, large injections made during period 4 contributes most to concentrations during hours 3 - 11, and those made during period 2 contributes from hours 15 - 1 (since the concentrations were periodic hours 0 and 24 are identical). Injections made during period 1 at nodes A and C had significant contributions

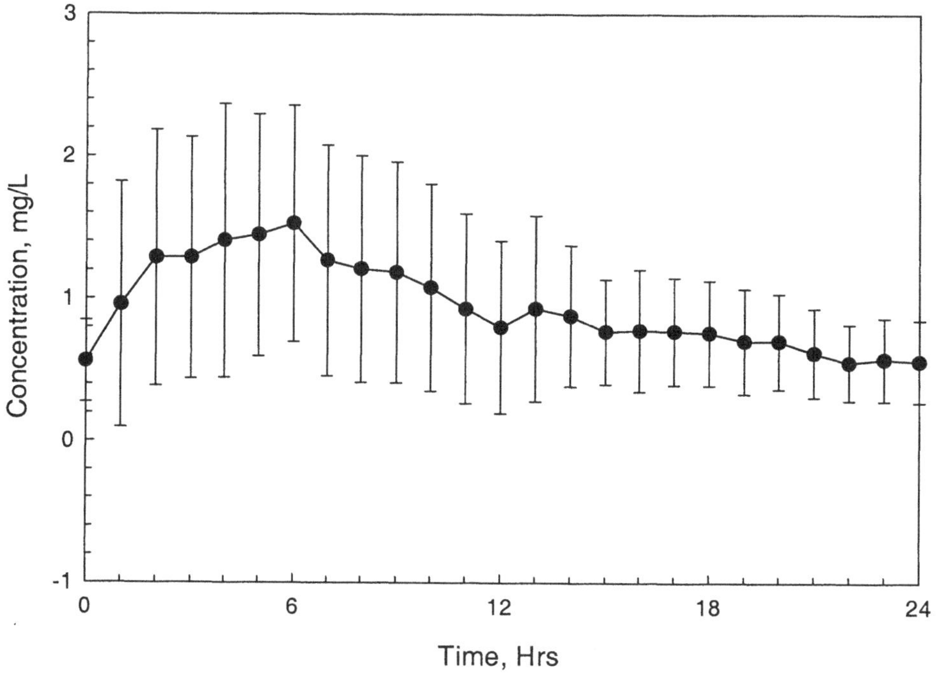

Figure 6.5: Average and standard deviation of concentrations, SCCRWA Case I.

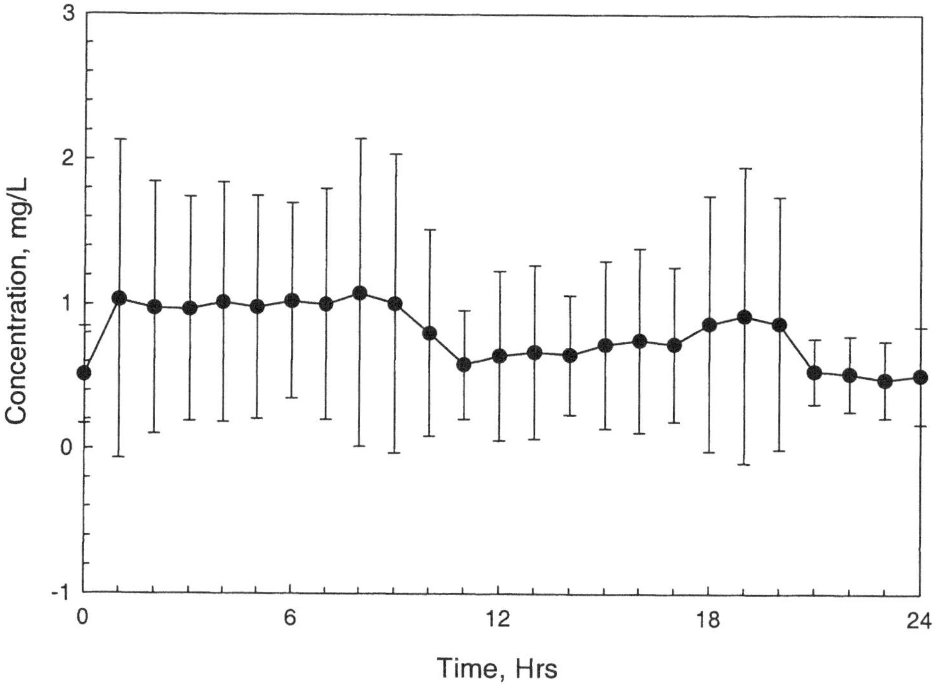

Figure 6.6: Average and standard deviation of concentrations, SCCRWA Case II.

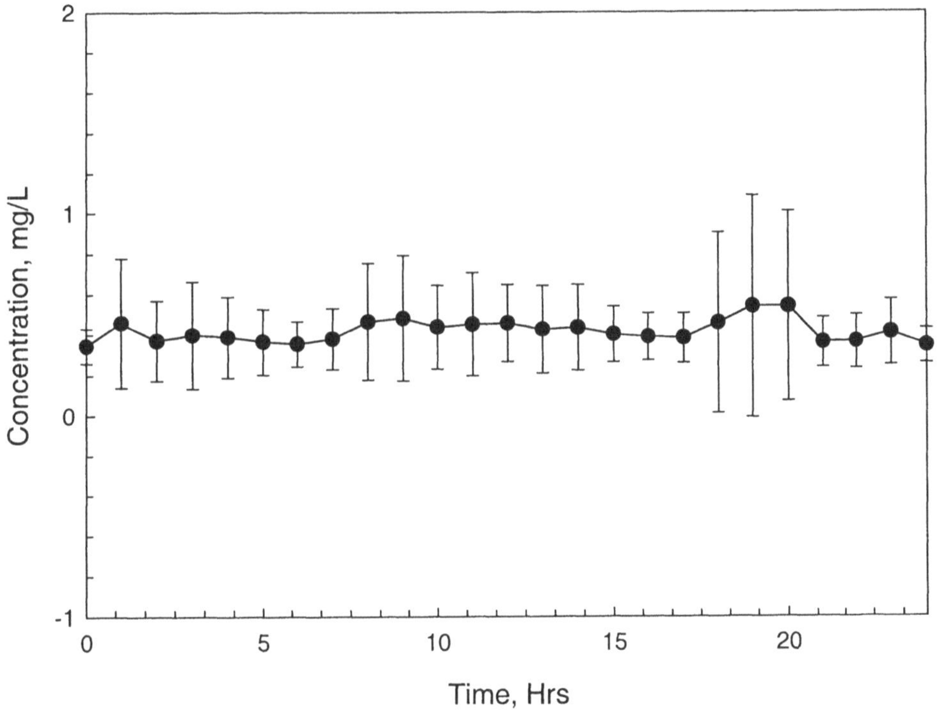

Figure 6.7: Average and standard deviation of concentrations, SCCRWA Case III.

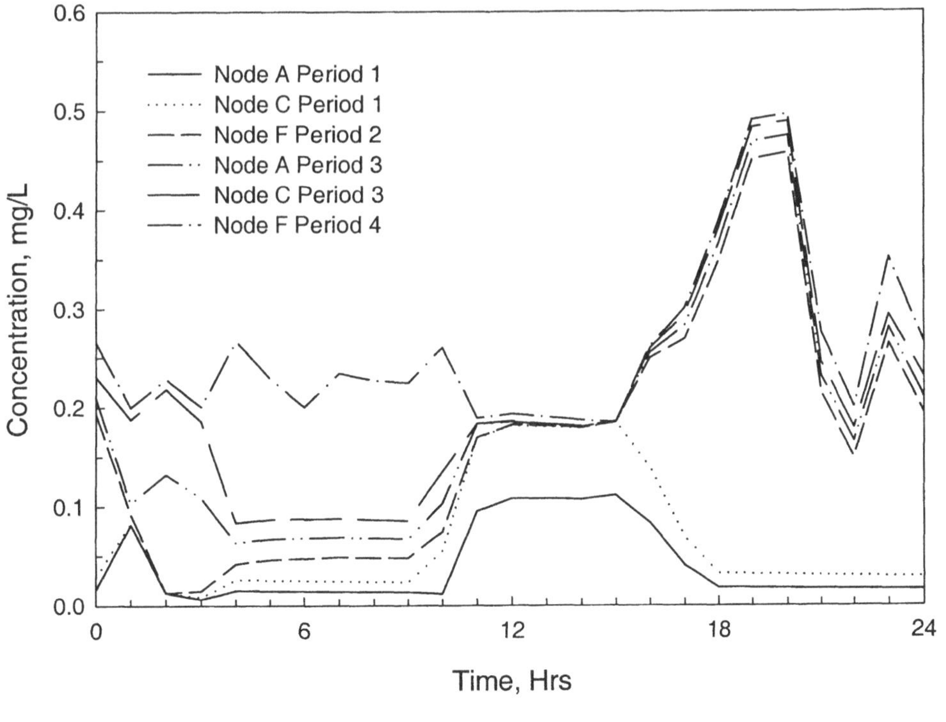

Figure 6.8: Individual booster station contributions to node 36, SCCRWA Case II.

during hours 10 - 18, and injections occurring during period 3 appeared from hours 0 - 4, revealing a 10 hour delay of the boost impacts. These delays occurred as a result of storage in the tank, and the transport time required to reach the observation node.

Since the concentration at any node is comprised of individual boosts arriving from many different locations and times, it is difficult to determine a priori the number and location of observation nodes whose concentrations would equal the minimum or maximum allowable concentration under an optimal dose schedule. The results of the simulations indicated, however, that the actual number of observation nodes where constraints were active was actually quite small. Out of 816 monitoring constraints (for each case), the schedule for case I was bounded by only 2 active constraints. The schedules for cases II and III were bounded by 7 and 15 active constraints respectively. The active lower constraints were generally located at dead-end nodes (e.g. nodes 10 and 36) or at the apex of loops (e.g. node 4). The active upper constraints were generally located adjacent to booster nodes (e.g. nodes 7 and 9).

While the design method includes dynamic hydraulics it does so with a qualification; these dynamics are required to be periodic over a cycle time T_h. As discussed in chapter 5, the finite-time formulation also assumed the mass injections to be periodic over a cycle time T_s. These two assumptions allow the concentrations c_j^m, and therefore the composite impulse response coefficients α_{ij}^{km}, to be periodic with cycle time T_α when $m \geq M$. Figure 6.9 shows the tank concentration dynamics resulting from implementation of the optimal schedule for Case I. The initial conditions of the system are overcome after t = 264 hours (11 days); thus M = 264 since Δt_m = 1 hour. Having already established that the network hydraulics are periodic (T_h = 24 hours), under the influence of periodic booster injections (T_s = 24 hours) the resulting disinfectant concentration dynamics are periodic as well (T_α = 24 hours). Therefore, the finite-time formulation will satisfy disinfectant concentrations for all times $m\Delta t_m \geq M\Delta t_m$, as anticipated.

6.1.4 Summary

The results indicate that booster disinfection can reduce the amount of disinfectant required to maintain an adequate residual throughout the distribution system, when compared to conventional disinfection at the source. The location of booster stations was shown to be important in maximizing booster disinfection efficiency. System hydraulics play an important part in determining the optimal schedule. In particular, a steady-state hydraulic approximation produced overly optimistic booster dose schedules, if tested against water quality dynamics that are more likely to represent actual operation. The optimal schedule is dependent, to some degree, on the tank concentration dynamics as shown by results for various booster locations. Among the five cases investigated, the best schedule was found when a booster station was located on the inlet/outlet line of a storage reservoir, eliminating the need to maintain an excessively large residual in the large volume of tank water, for distribution during high demand periods.

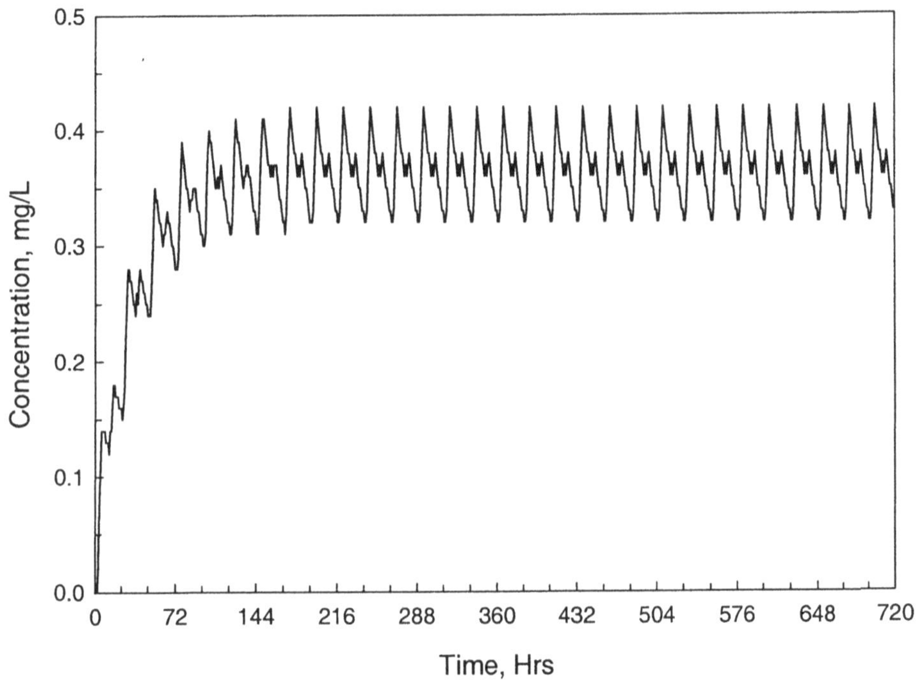

Figure 6.9: Concentration in tank resulting from optimal schedule for SCCRWA Case I.

Booster disinfection can allow a lower average disinfectant residual throughout the distribution system, and may also decrease the variability of disinfectant residual. The time variation of disinfectant concentration was also shown to be complex, as several different booster locations and intervals may be responsible for satisfying the residual constraints at any particular monitoring node.

A booster schedule could be obtained using existing water quality models, without the aid of mathematical optimization techniques provided by the BDDA software. Mass injection rates could be assumed for any set of booster locations and injection intervals, and the resulting simulated residuals evaluated to determine if they satisfy concentration constraints at the monitoring nodes. The mass injection rates could then be adjusted and the procedure repeated until a satisfactory boosting schedule is obtained. This iterative approach is, however, likely to be inefficient relative to the solution procedure described here. It could be difficult to identify feasible solutions using a trial and error technique, let alone those that approach the minimum disinfectant mass required. With some additional work to set up the network model properly, the linear programming formulation allows comparatively quick generation of alternative booster schedules, each of which can serve as the subject of further, more detailed, analysis.

6.2 South Norwalk Location/Scheduling Analysis

In the following sections, the BDDA optimal design methods are used to generate alternative booster disinfection designs for the gravity feed zone of the South Norwalk, Connecticut water distribution network. In contrast to the SCCRWA analysis, the booster station locations are not identified a priori, and the design problem is to determine simultaneously their best locations and mode of operation – defined as those locations and doses which minimize the total disinfectant mass applied while meeting residual maintenance constraints. These results also highlight the role of the disinfectant source type selection (`MASS`, `FLOWPACED`, `SETPOINT`), which mimics the booster station control mechanism and is shown to have a significant effect on results. The optimization results were obtained using the BDDA software and the `FORMULATION MINMASS` directive in the `[OPTIONS]` section of the design input file (see Appendix A).

The tradeoff between the total disinfectant mass required to maintain a minimum residual and the number of booster stations, as a function of their dosage schedules and locations, is generated as part of a comparison between conventional and optimal residual maintenance alternatives. The optimal designs produced show that the total disinfectant dosage applied to the network can be reduced while simultaneously improving residual maintenance coverage with the introduction of a modest number of booster stations. Further analysis quantifies reductions in the the mean and standard deviation of monitored residual concentrations, and in the total mass of disinfectant demanded (lost to bulk and wall decay reactions) within the network. As a result of addressing dosage reduction objectives explicitly, the designs generated may reduce disinfection by-product (DBP) concentrations at points of use, if reductions in disinfectant dosage lead to reductions in chlorine demand.

First, the network model and booster location model setup are described. Next, the optimal tradeoff between the number of booster stations built and average disinfectant mass dosages is evaluated using the model with a set of potential booster locations and varying operating options. Finally, an analysis will be used to predict the effect of the booster disinfection designs on residual maintenance in the network.

6.2.1 Network Model Description

The network model shown in Figure 6.10 includes 829 nodes, 3 tanks, and 1034 segments totaling 140.8 km (87.5 miles) of pipe. The total system demand is 19000 m^3/day (5.1 MGD), which is satisfied by gravity flow from the treatment plant located approximately 6 km (4 miles) north, north west, of the first tank. The modeled network hydraulics (user demands, pipe flows, and tank elevations) are periodic on a 24 hour cycle and are assumed to repeat indefinitely for planning purposes. The system hydraulic dynamics are shown in Figure 6.11, where flows into the system are negative and flows out of the system are positive. The diurnal demand cycle shown in the figure has two peaks. The late morning peak is the largest with a demand of approximately 23 m^3/min (6100 GPM); the early evening peak is slightly smaller in magnitude at 16 m^3/min (4200 GPM). The network has been designed to accommodate these peak flows, and requires little water from storage to do so. In fact, only 1100 m^3/day (0.3 MGD) of the total system demand (approximately 6%) is supplied from water in storage. Consequently, all three tanks exhibit shallow and irregular drain and fill cycles and remain full most of the time. The global bulk and wall decay coefficients used to model disinfectant decay reactions were $k_b = 0.53\ d^{-1}$ and $k_w = 5.1$ mm/day (0.20 in/day) respectively. Water quality modeling results, distribution system samples, and utility operating experience indicate that low and intermittent residual concentrations occur in the southern end of the network.

6.2.2 Booster Disinfection Design and Analysis (BDDA) Model Setup

The location and operation of injectors, and the locations and concentration limits at monitoring nodes are parameterized within the BDDA location model setup. The location model can be used to evaluate existing locations, such as disinfectant addition at the water treatment plant (WTP) or booster stations already built, and potential locations (through the `EXISTING` and `POTENTIAL` keyworks in the `[INJECTIONS]` section of the BDDA design input file – see Appendix A). The 30 booster locations (n_b= 30; 1 existing and 29 potential) are shown in Figure 6.10. Potential locations were chosen based on network hydraulics and engineering judgement. Some locations were selected near storage tanks and areas experiencing low or intermittent residual coverage, while others were selected to disperse the potential injectors throughout the network. The same monitoring constraints were used in all the location model runs, and require that residual concentrations be maintained between a minimum concentration, l_j= 0.20 mg/l, and a maximum concentration, u_j= 4.0 mg/l, over a 24 hour monitoring time window discretized using a one hour monitoring time step. (n_α= 24, resulting in hourly residual concentration checks.) The duration of the water quality simulations was conservatively set at 1008 hours with the monitoring time window starting time $M\Delta t_m$= 984 hours.

Monitoring at all demand nodes corresponds to the ideal of complete residual maintenance coverage. Water quality simulations indicated, however, that residual requirements could not be satisfied at some demand nodes even when all 30 booster stations operate at their maximum dose rates (given upper concentration bounds and the hydraulics occurring at each station). Thus, a strategy was required to prune nodes from the set of monitoring nodes for which constraints will be written (referred to as the monitoring constraint set). We believe that this step will be required for most network applications. The need to prune monitoring nodes may be indicative of real residual maintenance difficulties or an artifact of the water quality model developed for the network. There may, in fact, be locations in many networks that will experience low residuals even under a very aggressive booster disinfection strategy, and these may need to be managed by other means such as looping of dead end

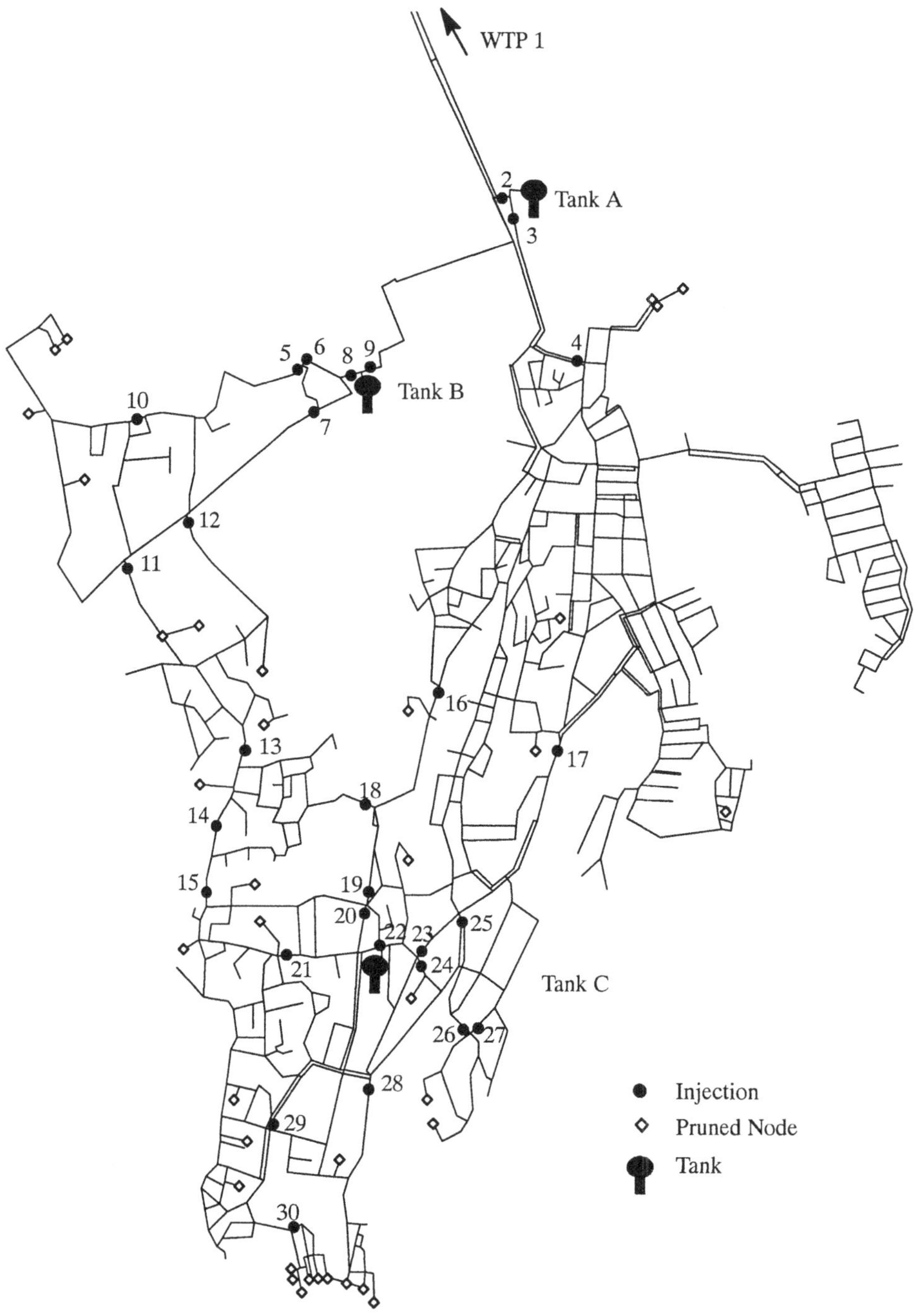

Figure 6.10: South Norwalk network map with potential locations and pruned nodes.

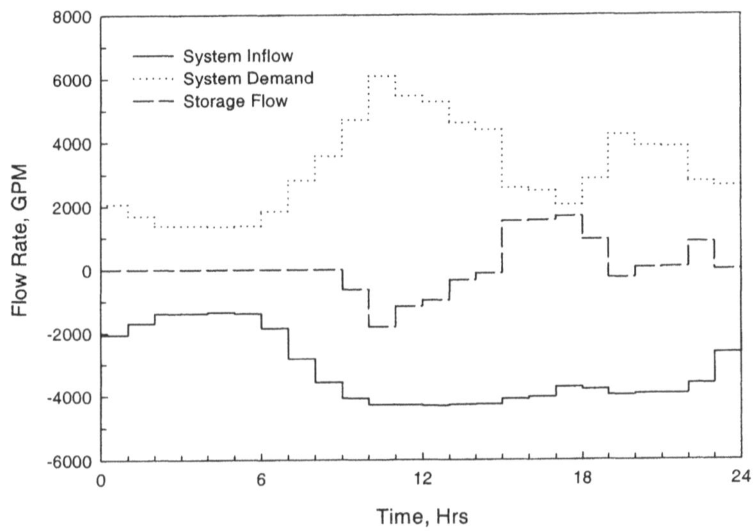

Figure 6.11: South Norwalk network hydraulics.

mains, or frequent and periodic flushing. Or, the low residuals could be an artifact of a model that includes the smallest water mains, where inaccurate demand estimates lead to large residence times. A network model may also include dead-end mains with zero demand (perhaps a location for future expansion), in which case the water age is infinite and minimum residuals are not required. In any of these or similar cases, a separate analysis (such as that described below) will be required to identify the set of nodes to exclude from monitoring.

A water age analysis was used to rank nodes and prune them from the monitoring constraint set in a meaningful way. Starting with all nodes in the network, the monitoring constraint set was constructed by removing nodes with residence times greater than the 95th percentile value of 50.8 hours (50th percentile - 15.6 hours, 100th percentile - 171.5 hours). The 36 nodes removed using this procedure are shown in Figure 6.10. The three tanks in the system, however, are included in the monitoring constraint set and thus, residual maintenance requirements were enforced there.

6.2.3 Results and Discussion

Optimal Tradeoff Analysis

The optimal designs discussed in this section examine the effect of dosage type and discretization on the tradeoff between the average disinfectant dosage rate and the number of booster stations. Table 6.3 shows the design cases that are examined. Cases 1 and 2 compare a single time interval schedule (n_i= 1) to a four time interval schedule (n_i= 4) for constant dosage rates during those intervals (`MASS` source type), and Cases 3 and 4 are the same as Cases 1 and 2 except that the dosages at the injectors are flow proportional (`FLOW` source

Table 6.3: South Norwalk Design Cases.

Case	Base Dose Variation	No. Dose Schedule Time Intervals	Interval Duration (hrs)
1	Constant	1	24
2	Constant	4	6
3	Flow Proportional	1	24
4	Flow Proportional	4	6

type). Cases 1 and 3 and 2 and 4 will be compared to illustrate the effect of this particular type of time varying dosage rate.

The constraint method of multiobjective programming(Cohon, 1978) was used to approximate the noninferior set of booster designs (the optimal tradeoff between the minimum average mass dosage rate and the number of booster station locations). The tradeoff solution sets were generated for values of n_b^{max} ranging from 3 booster stations (the minimum feasible number) to 20, and are shown in Figure 6.12 for Cases 2 – 4. Case 1 is not shown since all possible designs using the constant dosage rate with one scheduling time interval were infeasible (i.e. the monitoring constraints could not be satisfied for any set of constant doses).

By making explicit the optimal tradeoffs, the booster facility location model allows direct and logical comparison of booster system designs that are equivalent with regard to residual maintenance performance. The optimal tradeoffs indicate that the total average mass dosage rate depends not only on the number of sources but on how those sources are operated as well. Note the general shape of the tradeoffs in Figure 6.12 – the total dosage rate decreases significantly as the first few booster stations are added, after which the marginal improvement in total dosage rate per booster station diminishes. Case 3, using a single flow proportional dosage rate, requires the largest average dosage rate for any number of booster stations, with a minimum rate of approximately 24 kg/d. Case 2 requires less disinfectant mass by allowing 4 dosage rate variations per booster station, although the dosage rates are constant over their time interval; this case allows a minimum dosage rate of approximately 21 kg/d. Case 4 requires less disinfectant mass still (a minimum of approximately 19 kg/d), by allowing each of 4 dosage rates at each booster station to vary in proportion to the flow.

These differences among the design cases can be explained intuitively in terms of the operating configurations they represent. The ability of a design to reduce the average mass rate is related to the flexibility with which each station can adjust its input to match the flow rates and residence times within its residual coverage area. Comparing Cases 1 with 3, and 2 with 4, shows that the modulating effect of a flow proportional dosage rate on booster output concentration can be important, in terms of an increase in operational flexibility that allows for better tailoring of booster dose rates to network flows and residence times. Indeed,

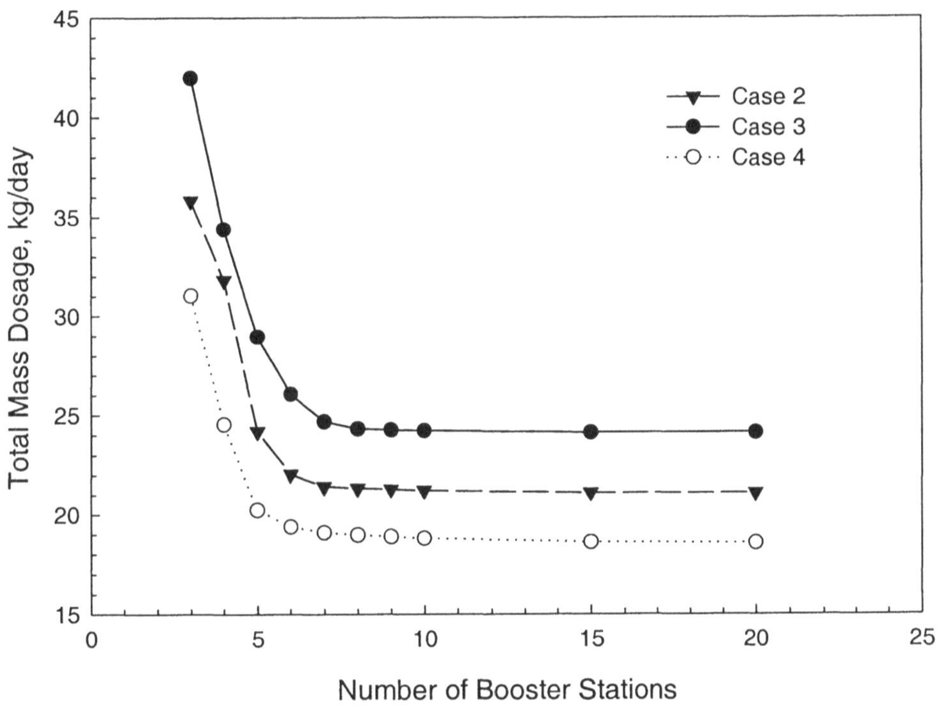

Figure 6.12: Location tradeoff curves for design cases.

the constant dosage rate in Case 1 was unable to maintain residuals without operating at a concentration in violation of the upper limit of 4.0 mg/L. These concentration violations are caused by variations in flow occurring at the injector, suggesting that booster stations operating with such a simple control scheme would be difficult to operate given the hydraulics occurring at the injector locations.

For both constant and flow proportional dosage rates, as the number of scheduling time intervals increased (from 1, for Cases 1 and 3, to 4, for Cases 2 and 4), the minimum average mass rate decreased. Similarly, this reduction in mass rate is attributed to the greater operating flexibility from the increased number of scheduling intervals. The reduction in average mass rate made possible through booster operating flexibility (allowing time variation or increasing the number of scheduling periods) is significant. Comparing the 3 station designs for Cases 3 and 4, increasing the number of schedule intervals from 1 to 4 reduced the average dosage rate by approximately 11 kg/d, nearly equaling the further reduction made possible by increasing the number of booster stations in Case 4. Thus, the dosage reduction potential of the designs was related to the control and scheduling decisions, as well as the number and location of booster stations.

Table 6.4: South Norwalk Optimal dosage schedules, x_i^k, for case 3 designs.

	Case 3 Design			
Node ID	**1 Station**	**3 Station**	**5 Station**	**7 Station**
1	4.00	2.30	1.23	0.94
2	-	-	0.93	1.18
9	-	-	0.41	0.64
8	-	0.48	0.54	0.52
16	-	-	-	0.04
25	-	-	-	0.13
22	-	1.08	1.74	1.17

Case 3 Booster Designs

In the following sections, the Case 3 designs developed previously are examined in greater detail. Case 3 operating conditions were selected since flow proportional control (a single time varying dosage rate proportional to the flow through the booster station) is a widely accepted practice for regulation of disinfectant dosages. Three designs were selected from the Case 3 solutions: the minimum booster station design (3 stations), the efficient minimum mass rate design (7 stations - corresponding to the sharp change in the tradeoff slope), and an intermediate design (5 stations). These designs will also be compared to a conventional residual maintenance scenario (1 station - without booster disinfection). The 1 station design represents the practice of applying all disinfectant within the treatment plant (the WTP outlet is Station 1). The 1 station design uses a flow proportional controller to maintain the maximum concentration of 4.0 mg/l entering the distribution system (the assumed disinfectant taste and odor threshold for residual maintenance operation, or the chlorine MCL); still, some monitoring set lower bound constraints are violated. Thus, the 1 station design *does not meet the same residual maintenance standards* as the other designs, and therefore is not a member of the noninferior tradeoff solution set; it is included as a reference for the purpose of comparison with a typical residual maintenance strategy.

The operating schedules for the designs are shown in Table 6.4. The values shown are the optimal dosage multipliers x_i^k, for a flow proportional dosage (see eq. (5.3). Referring to Table 6.4 and Figure 6.10, the 3, 5, and 7 station designs share many locations in common. The 3 station design locates stations at the outlet of the treatment plant (Station 1), and at the outlets of storage Tanks B (Station 8) and C (Station 22). The 5 station design shares these locations, and adds booster stations at the outlet of Tank A (Station 2) and at the inlet of Tank B (Station 9). To these, the 7 station design adds Station 16 north by north east of Tank C, and Station 25 east by north east of Tank C.

As mentioned previously, Station 1 is located at the entry into the distribution system. Its optimal operating schedule is the target concentration leaving the clearwell of the treatment plant, or the location of the first consumer. Thus, the potential exists for conflict between disinfection contact time and the dosage reduction objectives included in the location model. The concentrations added for each of the designs (see Table 6.4) show that the concentration entering the distribution system (Station 1) decreased from 4.0 mg/l for the 1 station design to 2.3 mg/l for the 3 station design, and to 0.94 mg/l for the 7 station. Experimental evidence suggests that DBP formation is a function of disinfectant demand (see section 4.7.3, also (Fleischacker and Randtke, 1983; Symons et al., 1996a; Stevens et al., 1989)), which may decrease with a reduction in disinfectant dose. Thus, significant reductions made to disinfectant dosages applied at the treatment plant would likely be accompanied by a reduction in DBP formation, at least until the booster dose. Disinfection requirements, however, may limit the ability to reduce doses at the plant. For example, the WTP in the application typically operates in the range of 1.00 to 1.20 mg/l which just meets disinfection requirements given their clearwell configuration and water quality. When disinfection requirements limit the solution (as they do for the 7 station design), specific alternatives include decoupling residual maintenance from disinfection requirements by introducing dechlorination after sufficient contact time(Nguyen and Montiel, 1994b), changing clearwell configuration, or specifying a minimum concentration at Station 1 (within the BDDA `[CONSTRAINTS]` section) to ensure that disinfection requirements are satisfied.

The decrease in disinfectant concentration leaving the plant is a result of disinfectant additions being redistributed to booster injectors located in the network. The extent of dosage redistribution is illustrated in Figure 6.13 for the 5 station design. The figure shows the average disinfectant concentrations at each node resulting from the 5 station design normalized by that for the 1 station design. In the region from the WTP extending down the center of the network, concentrations for the 5 station design are roughly a third (0.30 to 0.40) of those for the 1 station design. The magnitude of the average concentration differences was 2.77 to 1.63 mg/l less in the region extending from the plant. In contrast, the residuals in the lower portion of the network are on average 1 to 3 times that for the 1 station design, and the magnitude of the average residual differences ranged from 0.0 to 1.4 mg/l greater. Thus, concentrations have been redistributed so that they are lower where they had been needlessly high (near the treatment plant and the upper regions of the network), and higher where they had been too low (in the southern end of the network known to experience residual maintenance problems). As a result, customers would receive water with a concentration that is more closely tailored to their location.

As illustrated above, the designs generated by the BDDA location model can reduce unnecessary disinfectant exposure. The designs, however, may require refinement to be practical. For example, consider locations 8, directly upstream of Tank B, and 9, directly downstream, selected in the 5 and 7 station designs. These locations are so near each other that it would be impractical and unnecessary to build two stations. Examination of the hydraulics near Tank B indicate that the disinfectant dosages are being split such that Station 9 affects the concentration within the tank during the fill cycle, while both Stations 8 and 9, and the concentration within the tank, control the concentration leaving the tank during its

Figure 6.13: Ratio of average residual concentrations for 5-station design to those for 1-station design.

drain cycle. This configuration separates the task of maintaining a minimal residual in the tank, from that of maintaining a minimum residual at distant locations that receive stored water. Such operational flexibility could be efficient if there is significant bulk disinfectant decay occurring within the tank, and also in the water that drains from it on the way to the consumer. Thus when specifying potential locations near tanks, the timing of fill and drain cycles should be considered. If needed, the hydraulic model could be modified to allow operation of one booster station only during the fill cycle, and another only during the drain cycle. Should the optimal solution include both these locations, a practical implementation would specify a single booster station with appropriate piping design.

The optimal location model can be used to generate alternatives as part of a larger design process that must consider unmodeled issues. To illustrate this, consider Stations 16 and 25 which add a relatively small concentration to their flows in the 7 station design. It may not be necessary, or useful, to build these stations. A designer would need to consider the capital, operations, and maintenance cost of these two stations, versus the increase in average dosage rates (and associated health consequences) if they were not built. A range of other unmodeled issues, such as operational flexibility, and the residual maintenance coverage tradeoff (which was hidden from consideration by requiring equivalent residual maintenance for all optimal design alternatives) should also be considered. What is the benefit of maintaining a minimum 0.20 mg/l residual at those nodes which would violate this constraint, if Stations 16 and 25 were not built? How would the performance of a modified design not belonging to the optimal tradeoff set compare to the 5 station design located on the tradeoff curve? The answers to such questions are certainly supported by the quantitative information provided by the booster facility location model, but their real answers - in terms of health benefits - must also include quantitative analysis and subjective judgements made as part of a larger design process.

Design Performance

The optimal designs themselves contain a wealth of quantitative information that can reinforce engineering judgement. In this section some useful quantitative measures relating to residual maintenance and dosage reduction performance are developed using the Case 3 designs. Using a network water quality simulator, disinfectant concentrations occurring at network nodes can easily be estimated. Thus, the residual maintenance performance of the designs can be judged (visually and quantitatively) by simulating their behavior. The BDDA software supports this step through export of an Epanet input file configured with the optimal solution (`[EPANET]` keyword in `[OPTIONS]` section), and through automatic calculation of residual statistics (`[RANALYSIS]` keyword in `[OPTIONS]` section).

Design performance can be measured by demand weighting the residual concentrations at the monitoring nodes. Constructing a histogram is a convenient way to plot the demand weighted observations. The volume of water delivered within an arbitrary histogram bin spanning the concentration range $a < c_j^m \leq b$ is

$$V^{a,b} = \sum_{j,m \in S^{a,b}} D_j^m \Delta t_m, \tag{6.1}$$

where $S^{a,b} = \{j, m | a < c_j^m \leq b\}$, and D_j^m $[L^3/T]$ is the demand at node j and monitoring time m. For the current example, all demand nodes are considered, *including those pruned* to create the monitoring node set. The height of a bin in the histogram is simply $V^{a,b}/V_T$, where V_T $[L^3]$ is the total volume delivered during the monitoring time window.

A simple transformation converts the histogram to a cumulative distribution of concentrations (computed by summing the normalized histogram bins defined above). Using this distribution one can visualize disinfectant concentration trends on a network wide basis. Figure 6.14 contains the cumulative distribution of concentrations for the 1, 3, 5, and 7 station designs. There appears to be a significant difference between the 1 station and the other distributions. The 1 station distribution has a gradually increasing slope, compared to the booster designs which have slopes that are nearly vertical between 0.60 and 1.00 mg/l. The steeply sloped portion of the distribution corresponds to the concentration range over which a large fraction of the water is being delivered. For example, 11% of water was delivered with a concentration $\leq$ 1.0 mg/l for the 1 station design, compared to 20% for the 3 station design and increasing to 82% for the 7 station design. Conversely, the 1 station design delivers 82% of the demand with a concentration $\leq$ 3.6 mg/l. Though the delivered concentrations for the 7 station case ranged to a maximum of nearly 2.25 mg/l, the majority of the water was delivered with a concentration of 1.00 mg/l or less. Thus, the distribution of residual concentrations for the 7 station design, and to a lesser extent the 3 and 5 station designs, was less variable and lower on average, compared to the 1 station design.

The cumulative distributions can also be used to gauge residual maintenance performance. With the lower bound concentration specified in the location model serving as the minimum desirable residual, the volume of water delivered to all nodes with a concentration < 0.20 mg/l was computed and is shown in Table 6.5. The 1 station design delivered over 69000 l/d with a concentration < 0.20 mg/l (only 1.5% of the total network demand of 1.70E+7 l/d). The volumes delivered <0.20 mg/l was significantly less for the 3, 5, and 7 station designs. Thus the failures in residual maintenance for all of the designs occurred for an exceptionally small volume of the water delivered. Curiously, the volume delivered < 0.20 mg/l increased as the number of stations increased from 3 to 7 stations (See Table 6.5). Recall that 5% of the network nodes were pruned from the monitoring node set used to develop the designs. Because the pruned nodes were not in the monitoring set, any disinfectant delivered to them is considered in excess and the optimal solution tends to eliminate it. Thus, the concentrations delivered to nodes outside the monitoring set tend to decrease below the lower bound on concentration as the number of booster stations increases and the ability to tailor doses to the network residence time and flow distribution improves. If this is troublesome, one strategy would be to include these 5% of nodes within the monitoring set,

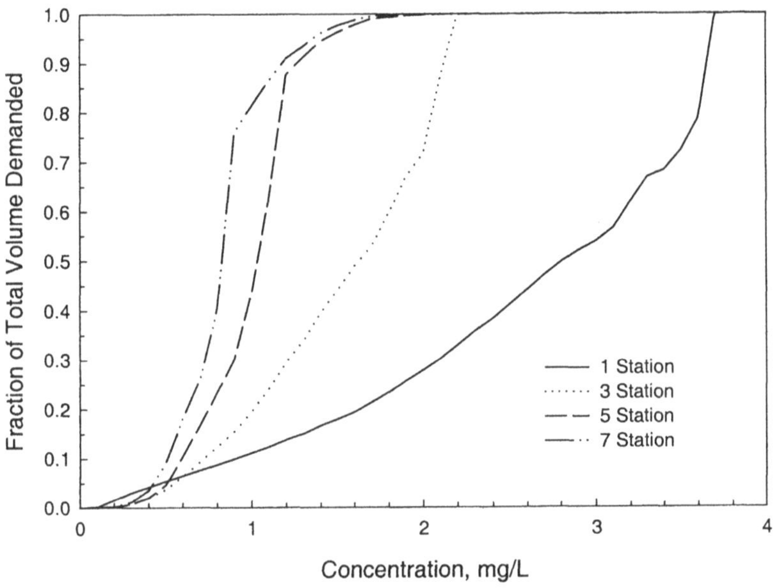

Figure 6.14: Cumulative distribution of concentrations for Case 3 designs.

but to allow a lower minimum concentration. One could then separately examine the trade-off between the residual concentration desired at "long residence time" nodes and number of booster stations or disinfectant mass required. Despite the low residual concentrations occurring at nodes outside the monitoring set, residual maintenance coverage is better for the 3, 5, and 7 station designs when compared to the 1 station design. Indeed, the minimum concentration delivered was 0.01 mg/l for the 1 station design, but ranged from 0.04 to 0.05 mg/l for all the other designs (minimum and maximum concentrations are shown in Table 6.5).

Descriptive statistics are easily generated from the distributions. The mean, standard deviation, and the coefficient of variation for each distribution are shown in Table 6.5. The mean concentration for the conventional design was 2.57 mg/l. As the number of stations increased the mean concentration decreased to 1.52 mg/l for the 3 station design and to 0.82 mg/l for the 7 station design. The standard deviation decreased from 1.04 mg/l for the 1 station design to 0.53 mg/l for the 3 station design, and to a low of 0.26 mg/l for the 7 station design. Thus, the statistics predict a decrease in the mean and standard deviation of the concentrations delivered and furthermore that these decreases are a function of the number of stations. This result is consistent with the intuitive notion that a greater number of booster stations allows more precise control over the spatial and temporal delivery of disinfectant.

A system wide mass balance on residual disinfectant can be performed using the dosage rate applied to network (the objective function of the optimal design model), and delivered by the network (derived from the distribution of concentrations), during the monitoring timc window (see Table 6.5). As the number of stations increased, the total mass applied was

Table 6.5: South Norwalk Case 3 design performance statistics, derived from concentration distribution.

	Case 3 Design			
Performance Statistic	**1 Station**	**3 Station**	**5 Station**	**7 Station**
Mean Conc. (mg/L)	2.57	1.52	0.98	0.82
Std. Dev. Conc. (mg/L)	1.04	0.53	0.27	0.26
Volume < 0.20 (mg/L)	69310	23887	31229	40333
Minimum Conc. (mg/L)	0.01	0.05	0.04	0.04
Maximum Conc. (mg/L)	4.00	2.44	3.16	2.25
Dosage Applied (kg/d)	69.70	42.00	29.00	24.70
Dosage Delivered (kg/d)	43.90	25.90	16.80	14.10
Dosage Demanded (kg/d)	25.80	16.10	12.20	10.60

reduced from 69.7 kg/day for the 1 station design to 42.0 kg/day for the 3 station, and to a low of 24.7 kg/day for the 7 station design. Similarly, as the number of stations increased the disinfectant mass delivered decreased from 43.9 kg/day for the 1 station design to 25.9 kg/day for the 3 station design, and to 14.1 kg/day for the 7 station design. Thus, the delivered dosages, and therefore, the exposure of consumers to disinfectant was significantly reduced for the booster station designs. Provided that the minimum residual is an adequate barrier to microbial contamination, this reduced exposure should be regarded as a public health benefit. It may also be of benefit to consumers that are sensitive to chlorine tastes and odors.

The mass balance can be taken a step further by using the dosages applied and delivered to estimate the dosage demanded within the network. A portion of the disinfectant mass consumed through bulk and wall decay reactions contributes to DBP formation. The results in chapter 4 (see also(Boccelli, 1999; Clark)) showed that first-order models may overestimate chlorine demand under conditions when oxidizable organic matter is limiting the reaction; thus first-order based estimates of dosage and demand reduction due to booster disinfection may be optimistic. Furthermore, the results in chapter 4 and the cited studies have shown that trihalomethane (THM) formation can be modeled as a linear function of chlorine demand, though little research is available to support this finding under rechlorination conditions in actual distribution systems. Thus, an optimistic estimate of the disinfectant demanded for the 1 station design was 25.8 kg/d and varied from 16.1 kg/d to 10.6 kg/d for the other designs. The percent reduction in disinfectant demand for the Case 3 booster designs ranged from 37.6% to 58.9% compared to the 1 station design. If THM formation is linearly related to disinfectant demand, the reductions estimated here will translate directly into public health benefits, though the actual size of this benefit will depend on the degree to which chlorine demand depends on chlorine concentration (see chapter 4).

6.2.4 Summary

Booster disinfection system designs were developed for an example network, and included simultaneous location and scheduling of the booster doses. The analysis quantified the tradeoff between the minimum disinfectant mass required to satisfy residual concentration constraints, and the minimum number of booster stations. Measures of design performance indicate that the disinfectant dosage required to satisfy residuals using booster disinfection is a function of injector station location and of their operation, including disinfectant source model (controller type and dose schedule discretization). A more detailed analysis on a family of designs for flow proportional booster stations indicate that significant disinfectant mass savings, compared to a conventional disinfection scenario, are possible. Furthermore, the booster station designs effectively redistributed disinfectant mass from the treatment plant into the distribution system, leading to a useful reduction in the mean and standard deviation of residual concentrations at demand nodes. Apparently, booster disinfection has the potential to reduce aggregate exposure of the population to chlorine (and potentially its by-products), while simultaneously improving residual coverage throughout the network.

6.3 Contra Costa Heuristic Location Analysis

In the following sections, the BDDA optimal design methods are used to generate alternative booster disinfection designs for the Contra Costa Water District (CCWD) distribution system. This application illustrates results obtained using the maximum set covering model formulation described in chapter 5. Unlike the previous minimum mass models, this formulation does not consider the scheduling of the booster doses; all booster doses are assumed to operate at the same constant set point concentration. The objective is to identify the minimize number of booster stations and their locations, given that the monitoring node concentrations meet a minimum standard. This formulation is intended mainly as a rough screening model, to identify a small set of nodes that may be good booster station locations. This formulation has two key advantages over the others: 1) it does not require formal optimization methods, and thus does not require the CPLEX commercial optimization solver (see Appendix A), 2) it scales well to very large networks, and in this application 1200 nodes were evaluated as potential booster locations.

The optimization results were obtained using the BDDA software and the `FORMULATION COVER` and `[HEURISTIC]` directives in the `[OPTIONS]` section of the design input file (the latter solves the optimization problem using a heuristic procedure rather than commercial optimization algorithms, see Appendix A). All potential booster stations were modeled as `SETPOINT` sources in the [INJECTIONS] section.

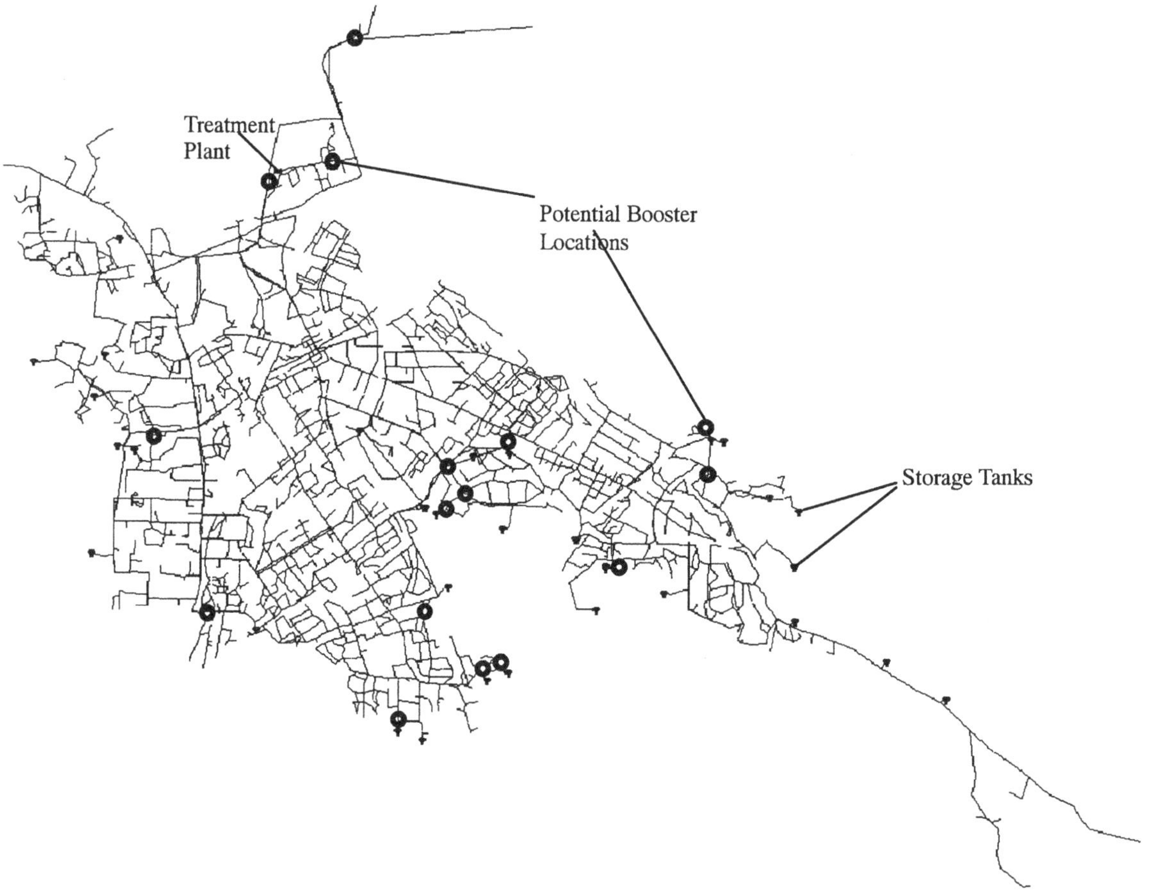

Figure 6.15: CCWD distribution system, with booster locations suggested by heuristic solution of maximum set covering model.

6.3.1 Results and Discussion

To illustrate briefly the type of results that can be obtained, the `COVER` model was used to locate candidate booster locations for the network shown in Figure 6.15; the model includes 3323 junctions, 3829 pipes, 1 reservoir, 34 tanks, 61 pumps, and 2 pressure control valves. All water is treated at a single source in the north, and is pumped in stages to successively higher service elevations. This chloraminated system can experience difficulty maintaining residual coverage in the southeast portion of the network, where average junction travel times can exceed three weeks.

The model was applied by selecting 1200 potential booster locations at random from among the 3323 junctions, and adding to these the 35 tank and reservoir junctions. The model was then solved for different values of the residual coverage parameter η (see chapter 5). As shown in Figure 6.16, as the ratio $\eta = c^{min}/c^{max}$ increases, the minimum number

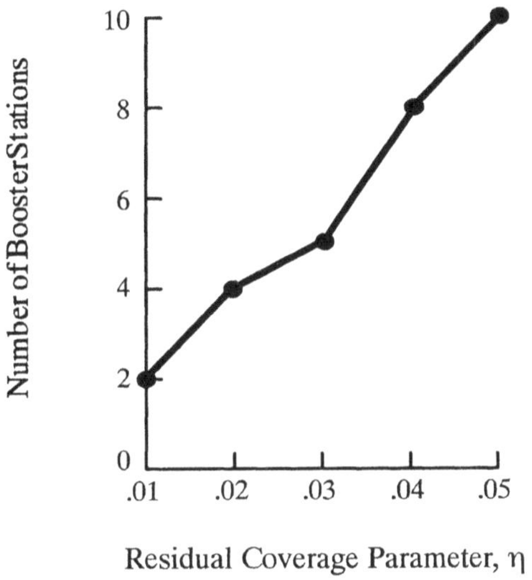

Figure 6.16: Minimum number of booster stations versus coverage parameter, η, for CCWD application.

of booster stations increases to accommodate the stricter definition of residual coverage. For each solution an injection station at the source is selected (of necessity), and other booster stations are selected at locations adjacent to tanks or at key junctions which deliver significant flow to critical parts of the network. Among all five designs in Fig. 6.16, a total of 16 different booster locations are selected. These locations are shown collectively in Fig. 6.15. The selected locations are geographically distributed throughout the network.

The results required approximately 12 hours of CPU time for problem setup on a 300 MHz Pentium II personal computer (mainly for calculation of impulse response coefficients). This setup time was required once for all five designs, as the impulse response coefficients are not affected by η, and the BDDA software allows reading from a binary file of saved impulse response coefficients. The solution time was on the order of several minutes; thus the vast majority of computation time is spent in problem setup. It is argued that these models are thus practicable for actual networks, and furthermore that the complexity of network dynamics would present a challenge for other non-systematic, model-based design procedures. Given the relatively small subset of selected locations, however, further analysis may be conducted in a more focussed manner using simulation, or these locations may be input to the `MINMASS` BDDA formulations for developing optimal dose schedules (in conjunction with other locations based on engineering judgement).

Appendix A

BDDA Software User's Guide

Booster Disinfection Design and Analysis (BDDA) User's Guide

Version 1.0

University of Cincinnati
Department of Civil and Environmental Eng.
Cincinnati, OH

A.1 Introduction

Booster disinfection is a residual maintenance technique that requires the selection of strategic locations in a network for reapplication of disinfectant. The process of booster disinfection design to this point has been one of trial and error. The Booster Disinfection Design and Analysis (BDDA) application described here will allow designers to develop booster station designs that minimize the number of stations and quantify other important treatment objectives including: 1) minimize the total mass dosage of disinfectant applied to a network; 2) minimize the maximum concentration monitored in a network; and 3) select the minimum set of stations necessary to achieve residual maintenance coverage requirements.

The BDDA is a MS-DOS console application based on the Epanet programmer's toolkit(Rossman, 1999), which is run from the command prompt on PC's running MS-Windows. This manual documents the input file syntax recognized by the program and takes the user step by step through an application using the program. The manual assumes prior knowledge of Epanet and it's input file format(Rossman, 2000).

A.2 BDDA Input File Organization

The BDDA input file is divided into four sections, each of which is denoted by a keyword in brackets.

Section	Description
`[INJECTIONS]`	Injector node identification and control options
`[CONSTRAINTS]`	Bounds on monitored concentrations, and the number of injector locations allowed to operate
`[OPTIONS]`	Miscellaneous analysis and output options

Section	Description
[SOLVER]	Optimization solver related options
[END]	End of input file

A.3 EPANET Input File Organization

Due to the nature of the Booster Disinfection Design and Analysis application and its operation with EPANET, several sections of the EPANET input file interact with the BDDA. The user should be aware of the particular interactions listed below. Please refer to the EPANET manual(Rossman, 2000) for more information about the EPANET input format.

Section	Interaction
[JUNCTIONS]	Negative demands are not allowed at injector locations
[SOURCES]	Data in the SOURCES section will be ignored
[QUALITY]	QUALITY section settings may affect optimization results
[OPTIONS]	Tolerance settings in OPTIONS must keep in mind the injection concentrations specified for the problem setup phase
[TIME]	Simulation duration determines period in which impact coefficients are written. Pattern time step setting in EPANET will also effect pattern time step used by the BDDA

A.4 BDDA Input File Format

This section describes the section and keywords recognized by the BDDA. Each section in the BDDA input file, as in EPANET, can contain one or more lines of data. Blank lines may appear anywhere in the file, and comments can be added by preceding them by a semicolon (;). Lines are limited to 80 characters or less. Table A.1 summarizes the BDDA default units and algorithmic settings, and Table A.2 recommends key parameter values and options for both Epanet and BDDA.

Table A.1: Summary of BDDA default units and settings

Parameter	Default Value	Relevant Section or Keyword
Injector Status	Ignore	`[INJECTIONS]`
Injector Type	Mass Source Type	`[INJECTIONS] TYPE`
Injector Schedule	None	`[INJECTIONS] SCHEDULE`
Monitoring Node Locations	None	`[CONSTRAINTS] CONC`
Minimum Concentration	-Infinity mg/L	`[CONSTRAINTS] L`
Maximum Concentration	+Infinity mg/L	`[CONSTRAINTS] U`
Number of Injectors Selected	Number of Injector Nodes	`[CONSTRAINTS] SELECT`
Max. Injection Rate Capacity	100 kg/day	`[CONSTRAINTS] MAXIRC`
Optimization Formulation	Minmass	`[OPTIONS] FORMULATION`
Influence Coefficient Binary File	Scratch File (not saved)	`[OPTIONS] INFLUENCE`
MPS Format File	None	`[OPTIONS] MPS`
EPANET Input File	None	`[OPTIONS] EPANET`
Minimum Impact Coefficient	1.0e-6	`[OPTIONS] MINCOEFF`
Perturbation Injection Dosage	500.0 mg/L	`[OPTIONS] INJDOSE`
Monitoring Period Duration	1 max. schedule cycle duration	`[OPTIONS] MONITOR`
Mass Avg. Period Duration	1 schedule cycle duration	`[OPTIONS] MASSAVG`
CPLEX Solution File	None	`[SOLVER] CPLEX`
CPLEX Solution Cutoff	1.00E+75	`[SOLVER] SLNCUTOFF`
CPLEX Optimality Tolerance	1.00E-04	`[SOLVER] SLNTOL`
CPLEX Tree File Options	0	`[SOLVER] TREEFILE`
CPLEX Time Limit	1.00E+75	`[SOLVER] TIMELIM`
CPLEX Tree Memory Limit	1.00E+75	`[SOLVER] TREEMEMLIM`

Table A.2: Summary of recommended EPANET and BDDA settings

Parameter	Recommended Setting	Relevant Section or Keyword
Nodal Demands at Injectors	Nodal demands ≥ 0	[JUNCTIONS],[INJECTIONS]
Injectors at Tanks	Avoid specifying injectors at tank nodes	[TANKS], [INJECTIONS]
Injector Types	Avoid mixing injector types	[INJECTIONS] TYPE
Minimum Monitoring Concentration	Monitor at all nodes for a minimum detectable concentration	[CONSTRAINTS] CONC L
Maximum Monitoring Concentration	Monitor at all injector nodes for a maximum allowable concentration	[CONSTRAINTS] CONC U
Source Quality	Sources section is ignored	[SOURCES]
Initial Quality	Initial quality = 0.0 mg/L for all nodes, tanks, and reservoirs	[QUALITY]
Simulation Duration	Minimum duration $> 10\times$ maximum schedule cycle duration	[TIMES] DURATION
Pattern Timestep	BDDA and EPANET use the same value	[TIMES] PATTERN TIMESTEP
Water Quality Simulation Tolerance	Perturbation dosage/400	[OPTIONS] TOLERANCE,[OPTIONS] INJDOSE
CPLEX Tree Memory Limit	Slightly less than the total available system memory	[SOLVER] TREEMEMLIM

A.4.1 [INJECTIONS]

Purpose

Specify existing and potential injector locations and source types, and also the schedules that determine how injections at each location vary over time.

Keywords

Keyword	Description
EXISTING	Specify existing injector location(s)
POTENTIAL	Specify potential injector location(s)
IGNORE	Ignore injector location during analysis
TYPE	Specify injection type for specified location(s)
SCHEDULE	Set injection scheduling pattern for specified location(s)

Format

<table>
<tr><th>Keyword</th><th>Parameters</th><th></th></tr>
<tr><td>EXISTING</td><td><<i>nodeID</i>|ALL[†]></td><td></td></tr>
<tr><td>POTENTIAL</td><td><<i>nodeID</i>|ALL[†]></td><td></td></tr>
<tr><td>IGNORE</td><td><<i>nodeID</i>></td><td></td></tr>
<tr><td>TYPE</td><td><<i>nodeID</i>|ALL[†]></td><td><MASS|SET|FLOW></td></tr>
<tr><td>SCHEDULE</td><td><<i>nodeID</i>|ALL[†]></td><td><<i>duration</i>> <<i>start1 stop1,...,startN stopN</i>></td></tr>
</table>

[†]When specifying injector options, the keyword **ALL** will override settings made to individual nodes specified previously using ***nodeID***. Thus any statements using the **ALL** keyword should precede node-specific statements.

Parameters

Parameter	Description
nodeID	Epanet Node ID string
ALL	Select all network nodes
duration	Injection scheduling cycle duration (no. pattern time steps)
starti stopi	Injection period i start/stop times (pattern time steps)
MASS	Use mass type source for injector
SET	Use setpoint type source for injector
FLOW	Use flowpace type source for injector

Example

```
[INJECTIONS]                          ;Start of section
POTENTIAL     ALL                     ;Consider all nodes in the system as
                                      ;potential locations
EXISTING      101                     ;Reset potential setting at Node 101
                                      ;to existing
IGNORE        536                     ;Overrides previous settings to ignore
                                      ;injection at Node 536
TYPE          ALL MASS                ;All injectors are mass sources
SCHEDULE      ALL 24 1 12 13 24       ;Injection schedule cycle 24 hours
                                      ;Two 12 hour periodic injections specified,
                                      ;assuming one hour pattern time step
```

A.4.2 [CONSTRAINTS]

Purpose

Set upper and lower concentration limits at monitoring node locations, and select the maximum number of potential injector locations allowed (for formulations having one or more POTENTIAL injector locations).

Keywords

Keyword	Description
CONC	Specify a concentration constraint at monitoring node location(s)
SELECT	Specify the number of injectors allowed to operate
IGNORE	Ignore monitoring node location during analysis
MAXIRC	Specify the maximum allowable injection rate capacity (kg/day)

Format

Keyword	Parameters
CONC	<*nodeID*|ALL[†]> <L|U> <*cvalue*>
SELECT	<*value*>
IGNORE	<*nodeID*>
MAXIRC	<*value*>

[†] When specifying constraint options, the keyword ALL will override settings made to individual nodes specified previously using *nodeID*. Thus any statements using the ALL keyword should precede node-specific statements.

Parameters

Parameter	Description
nodeID	Epanet Node ID string
ALL	Select all network nodes
L	Specify a lower (minimum) concentration limit
U	Specify an upper (maximum) concentration limit
cvalue	Allowable disinfectant concentration (mg/L)
value	Numerical value

Example

```
[CONSTRAINTS]                      ;Start of section
CONC          ALL   L  0.20        ;Lower limit of 0.20 mg/L set for
                                   ;all nodes
CONC          101   L  0.00        ;Reset lower limit for node ID 101
CONC          ALL   U  4.00        ;Upper limit of 4.00 mg/L set for
                                   ;all nodes
IGNORE        26                   ;Override constraint settings to
                                   ;ignore node 26
SELECT        2                    ;Select a maximum of two out of all
                                   ;potential injector locations
MAXIRC        23.0                 ;Allow a total of 23 kg/day to be
                                   ;applied at all injector locations
```

A.4.3 [OPTIONS]

Purpose

Specify the optimal design formulation, influence coefficient binary file, optimization output format, and miscellaneous algorithmic options and values.

Keywords

Keyword	Description
FORMULATION	Specify an optimization formulation type
HEURISTIC	Approximate solution of certain optimization formulations
INFLUENCE	Specify binary influence coefficient file and status
CHECK	Perform various automatic checks
MPS	Write standard MPS format file for exporting optimization problem
EPANET	Write Epanet input file with injector scheduling data
RANALYSIS	Analyze disinfectant residual maintenance performance
RERROR	
MINCOEFF	Specify minimum influence coefficient
MINDEMAND	Minimum demand at monitoring nodes (depends on CHECK)
MINFLOW	Minimum flow at injector nodes (depends on CHECK)
INJDOSE	Specify injector concentration used during perturbation analysis
MONITOR	Specify monitoring period length (number of schedule cycles)
MASSAVG	Specify period over which to compute impact coefficient normalization constant
DURTOL	

Format

<table>
<tr><th>Keyword</th><th>Parameters</th><th></th><th></th><th></th></tr>
<tr><td>FORMULATION</td><td><MINMASS|MINCONC|COVER></td><td></td><td></td><td></td></tr>
<tr><td>HEURISTIC</td><td></td><td></td><td></td><td></td></tr>
<tr><td>INFLUENCE</td><td><SAVE|USE></td><td><<i>infile/outfile</i>></td><td></td><td></td></tr>
<tr><td>CHECK</td><td><MONITOR></td><td><INFEAS1|INFEAS2|AVGDEMAND></td><td></td><td></td></tr>
<tr><td>CHECK</td><td><SIMULATION></td><td><DURATION></td><td></td><td></td></tr>
<tr><td>CHECK</td><td><INJECTOR></td><td><FLOWOUT|FLOWREVERSE></td><td></td><td></td></tr>
<tr><td>MPS</td><td><<i>outfile</i>></td><td></td><td></td><td></td></tr>
<tr><td>EPANET</td><td><<i>outfile</i>></td><td></td><td></td><td></td></tr>
<tr><td>RANALYSIS</td><td><<i>infile</i>></td><td><<i>outfile</i>></td><td><<i>nperiods</i>></td><td><<i>nbins</i>></td></tr>
<tr><td>RERROR</td><td><<i>value</i>></td><td></td><td></td><td></td></tr>
<tr><td>MINCOEFF</td><td><<i>value</i>></td><td></td><td></td><td></td></tr>
<tr><td>MINDEMAND</td><td><<i>value</i>></td><td></td><td></td><td></td></tr>
</table>

Keyword	Parameters
MINFLOW	<*value*>
INJDOSE	<*value*>
MONITOR	<*value*>
MASSAVG	<*value*>
DURTOL	<*value*>

Parameters

Parameter	Description
MINMASS	Minimizes total mass dosage of disinfectant applied to the network through all injector locations
MINCONC	Minimizes maximum disinfectant concentration at all injector locations
COVER	Specify the maximum set covering location formulation
SAVE	Calculate and save binary influence coefficient file
USE	Use previously saved binary influence coefficient file
MONITOR	Check/modify monitoring locations based on feasibility or demand
INJECTOR	Check/modify injector locations based on hydraulic characteristics
SIMULATION	Check water quality simulation
INFEAS1	
INFEAS2	
AVGDEMAND	Eliminate monitoring requirements at nodes with demand less than MINDEMAND
FLOWOUT	Eliminate injector locations at nodes with total outflow less than MINFLOW
FLOWREVERSE	Eliminate injector locations at nodes with flow reversals
DURATION	Check simulation duration to insure stationarity of concentration dynamics (periodicity)
infile	Name of input disk file
outfile	Name of output disk file
nperiods	
nbins	
value	Numerical value

Example

```
[OPTIONS]                                  ;Start of section
FORMULATION  COVER                         ;Maximum set covering location
                                           ;problem
INFLUENCE    SAVE        net2vv.inf        ;Save influence coefficients to
                                           ;net2vv.inf
MPS          net2vv.mps                    ;Save an MPS format input file for
                                           ;external solver
MINCOEFF     1.0e-6                        ;Influence coefficients < 10^6 are
                                           ;ignored
INJDOSE      500.0                         ;Use concentration of 500 mg/L for
                                           ;perturbation analysis phase
MONITOR      2                             ;Force monitoring period to be two
                                           ;times the length of the maximum
                                           ;injection schedule
CHECK        DURATION                      ;Report estimate of maximum
                                           ;percentage error in impact
                                           ;coefficient periodicity
```

A.4.4 [SOLVER]

Purpose

Specify parameters that affect performance of the CPLEX optimization algorithms.

Keywords

Keyword	Description
CPLEX	Solve optimization problem using CPLEX and write a solution file
SLNCUTOFF	Specify upper bound for the optimal solution
SLNTOL	Specify solution optimality tolerance (%)
TREEFILE	Specify search tree file options
TIMELIM	Specify maximum run time for optimization procedure (sec)
TREEMEMLIM	Specify maximum search tree size (Mb)

Format

Keyword	Parameters
CPLEX	<*outfile*>
SLNCUTOFF	<*value*>
SLNTOL	<*value*>
TREEFILE	<0|1|2|3>
TIMELIM	*value*
TREEMEMLIM	*value*

Parameters

Parameter	Description
outfile	Name of output disk file
value	Numerical value
0	No tree file
1	Store tree file in memory and allow compression
2	Store tree file on disk
3	Store tree file on disk and allow compression

Example

```
[SOLVER]                        ;Start of section
CPLEX        net2vv.txt         ;Save optimal results to net2vv.txt
SLNCUTOFF    30.0               ;Ignore solutions applying > 30kg/day
TREEFILE     2                  ;Store tree file on disk
TREEMEMLIM   60                 ;Store tree file when memory usage > 60Mb
```

A.5 Developing Booster Station Designs

Application of the BDDA begins with the development of an EPANET input file containing the description of the network topographic information. Hydraulic and water quality calibration of a network model is necessary, as the solutions produced by the BDDA will be a function of hydraulics, residence times, and the bulk and wall disinfectant decay characteristics of the network. Discussion of these steps is beyond the scope of this manual, yet, they compose the most important prerequisite to applying the method illustrated here. Therefore, the BDDA should be considered an advanced application requiring the development and calibration of a network model and that users have practical experience with, and a theoretical understanding of, hydraulic and water quality simulation methods.

The BDDA is an optimization-based method using linear and mixed integer linear programming techniques. The application has been developed so that little experience with optimization methods is required of the user, although some exposure to techniques and model formulations would be helpful. After working through the examples in this section you should become comfortable enough with the optimization techniques to productively use the BDDA.

In the following sections, the steps recommended when performing an analysis using the BDDA will be illustrated for the simple "Net2" example network provided with the base Epanet software distribution. A network schematic is shown in figure 6.1. Detailed results from this application are presented in chapter 6, section 6.1.

1. Preparation of calibrated EPANET input file
2. Preparation of the BDDA input file
3. Solving optimization problems and interpreting solutions

A.5.1 Preparing the EPANET input file

Essentially, the BDDA calls EPANET in batch mode, running many water quality simulations, and organizes the data that results in a useful format. Thus, the EPANET input file must contain all of the sections necessary to run an extended period water quality simulation. The BDDA has been designed to require little or no modification to EPANET input files once they have been prepared, however, there are some potential interactions between EPANET and the BDDA that require some attention to detail. In this section we will review the format of a calibrated EPANET input file paying special attention to the modifications necessary to run an analysis using the BDDA. The Epanet input file to be discussed below is listed in Table A.3.

Table A.3: Example Epanet input file for "NET2"

```
[TITLE]
EPANET Example Network 2
[JUNCTIONS]
;-------------------------------
;       Elev.      Demand     Demand
;  ID   ft.        gpm        Pattern
;-------------------------------
   2    100        8
   3    60         14
   4    60         8
   5    100        8
   6    125        5
   7    160        4
   8    110        9
   9    180        14
   10   130        5
   11   185        34.79
   12   210        16
   13   210        2
   14   200        2
   15   190        2
   16   150        20
   17   180        20
   18   100        20
   19   150        5
   20   170        19
   21   150        16
   22   200        10
   23   230        8
   24   190        11
   25   230        6
   27   130        8
   28   110        0
   29   110        7
   30   130        3
   31   190        17
   32   110        17
   33   180        1.5
   34   190        1.5
   35   110        0
```

Table A.3: Example Epanet input file for "NET2"

```
   36     110        1
   71     160        0
   72     160        0
   291    110        0
   292    110        0
   101    50         0
[TANKS]
;-----------------------------------------
;         Elev.      Init.     Min.     Max.    Diam.
;  ID     ft.        Level     Level    Level   ft.
;-----------------------------------------
   26     235        56.7      50       70      50
[QUALITY]
;-------------------------
;  First  Last       Chlorine
;  Node   Node       mg/L
;-------------------------
   1      292        0.0
[SOURCES]
;---------------------------
;         Chlorine   Source
;  ID     mg/L       Pattern
;---------------------------
   1      1.0        3
[PIPES]
;-----------------------------------------
;         Head       Tail      Length   Diam.   Rough.
;  ID     Node       Node      ft.      in.     Coeff.
;-----------------------------------------
   1      1          101       1        12      100
   2      2          5         800      12      100
   3      2          3         1300     8       100
   4      3          4         1200     8       100
   5      4          5         1000     12      100
   6      5          6         1200     12      100
   7      6          7         2700     12      100
   8      7          71        1        12      140
   9      7          72        1        12      100
   10     8          10        1000     8       140
   11     9          11        700      12      100
```

Table A.3: Example Epanet input file for "NET2"

```
  12    11    12    1900   12   100
  13    12    13    600    12   100
  14    13    14    400    12   100
  15    14    15    300    12   100
  16    13    16    1500   8    100
  17    15    17    1500   8    100
  18    16    17    600    8    100
  19    17    18    700    12   100
  20    18    32    350    12   100
  21    16    19    1400   8    100
  22    14    20    1100   12   100
  23    20    21    1300   8    100
  24    21    22    1300   8    100
  25    20    22    1300   8    100
  26    24    23    600    12   100
  27    15    24    250    12   100
  28    23    25    300    12   100
  29    25    26    100    12   100
  30    25    31    600    12   100
  31    31    27    400    8    100
  32    27    29    400    8    100
  34    29    292   1      8    100
  35    22    33    1000   8    100
  36    33    34    400    8    100
  37    32    19    500    8    100
  38    29    291   1      8    100
  39    35    30    1000   8    100
  40    28    35    700    8    100
  41    28    36    300    8    100
  101   71    8     1199   12   140
  102   72    9     399    12   100
  105   292   28    699    8    100
  106   291   35    499    8    100
  107   101   2     2399   12   100
[REACTIONS]
Global Bulk -0.5
[PATTERNS]
;Demand pattern
1  1.19   0.97    0.90    0.90    0.82   1.12
1  1.21   0.60    0.60    1.27    2.39   0.90
```

Table A.3: Example Epanet input file for "NET2"

```
1  0.85   0.61      1.36      0.54      0.24   0.71
1  0.30   0.60      1.19      1.49      1.12   1.16
;Pump flow pattern
2 0.96 0.96 0.96 0.96 0.96 0.96 0 0 0 0 0 0
2 0.80 1.00 1.00 1.00 1.00 0.15 0 0 0 0 0 0
;Chlorine source pattern
3 1 1 1 1 1 1 1 1 1 1 1 1
3 1 1 1 1 1 1 1 1 1 1 1 1
[REPORT]
node none
links none
[TIMES]
DURATION 360
[OPTIONS]
QUALITY Chlorine
MAP NET2.MAP
TOLERANCE 0.01
[END]
```

[JUNCTIONS]

Any network node can serve as an injector location as long as the base demand at the node is zero or positive. All points of disinfectant injection, including booster stations (nodes 71, 72, 291, 292, 261) and the source node (node 101), are nodes that have been added to the original network and assigned a zero demand. A BDDA error will occur if injector nodes have negative demands.

[TANKS]

Tanks and reservoirs have two significant affects on the BDDA analysis, 1) Hydraulic and water quality initial conditions will affect the hydraulic and water quality dynamics, and 2) special attention is required when evaluating them as potential injector locations. The initial tank elevation in the [TANKS] section has been selected so that it corresponds to the eventual cyclical behavior of the hydraulic dynamics under periodic water demands. While it is true that, in general, hydraulic initial conditions will be overcome more rapidly than concentration initial conditions, our purpose is to eliminate any period of transient hydraulic dynamics at the beginning of the simulation. This is not required, but the simulation duration must allow for hydraulic and water quality dynamics to become stationary; any strategy for reducing this time will allow a shorter simulation duration, and thus a more efficient solution.

Booster injectors can be located at tanks and reservoirs, however, the behavior of EPANET for such sources should be considered. Sources at tanks and reservoirs are modeled in EPANET such that the injection is applied to the flow volume *exiting* during a water quality time step, as opposed to the stored volume in the tank or reservoir. Thus, placing a source at a tank or reservoir is equivalent to placing one at each outlet flow, with the inflows being unaffected. It is wise to remember this behavior when preparing for or interpreting a BDDA analysis.

`[PIPES]`

In the [PIPES] section of the EPANET input file the nodes identified as booster stations have been integrated into the pipe network. Pipe links between nodes have been added, deleted, and subdivided as necessary to accommodate the addition of the booster stations (see links 1, 34, 38, 101, 102, 105, 106, 107). If nodes representing booster stations or monitoring locations are added to links in the EPANET model, the map file should also be updated to include these new locations for correct visualization of the network. If nodes have not been added to the EPANET input file modification of the [PIPES] section and the map file would not be needed.

`[QUALITY]`

In the [QUALITY] section the initial concentrations assumed for all network nodes should ordinarily be set equal to zero. The BDDA will use the same concentration initial conditions for all booster injection impact calculations, and these initial conditions will affect the minimum simulation duration required to calculate correct booster injection impacts. Setting the concentration initial conditions to zero for all nodes (including tanks and reservoirs) is generally the best approach, unless significant prior analysis of water quality dynamics shows that a particular set of initial conditions can reduce the minimum required simulation duration for *all* booster injections. As one example, consider a booster injection that would not ordinarily have any impact at a particular monitoring node and time. If non-zero initial concentrations are used, the simulation duration will need to allow the effect of these initial concentrations to dissipate, else a false impact would be recorded for the booster injection.

`[SOURCES]`

All data in the `[SOURCES]` section is *ignored* by the BDDA. The source locations and patterns are specified instead in the `[INJECTIONS]` section of the BDDA input file. Thus the BDDA assumes complete control over all water quality sources. The source concentrations will be determined automatically by accounting for the flows exiting the injector locations.

`[REACTIONS]`

The solutions produced by the BDDA will be a function of the disinfectant decay parameters specified in the `[REACTIONS]` section of the Epanet input file. Water quality calibration of the network for analysis using the BDDA must include setting the bulk and wall reaction coefficients appropriately, as well as the important factors that determine network hydraulic dynamics (e.g. node demands, pipe and pump characteristics). The example ignores demand due to pipe wall material.

`[TIMES]`

The `[TIMES]` section of the Epanet input file affects both the total simulation duration used for each booster injection, and the pattern time step used by the BDDA. The simulation duration must be of sufficient length to allow a stationary concentration response at each network node, for each booster injection evaluated (a "stationary concentration response" can be interpreted as one where the daily average node concentration is not changing from one day to the next). We will also call this minimum required simulation duration the "equilibrium time." *The simulation duration is the most important simulation parameter in terms of its effect on accuracy of the BDDA results and its effect on BDDA computation time.* The longest equilibrium times will often be associated with concentrations in storage tanks due to disinfectant doses made at the water source(s). The equilibrium time can be estimated by making some exploratory Epanet runs and analyzing the concentration response, or by specifying `CHECK DURATION` in the BDDA `[OPTIONS]` section. The results of the `CHECK DURATION` option will be interpreted in a following section.

The `PATTERN TIMESTEP` affects the BDDA by determining the pattern time step used to specify injection patterns in the `[INJECTION]` section, and by determining the frequency that influence coefficients are written to the binary influence coefficient file.

`[OPTIONS]`

The `QUALITY` keyword must be specified in the `[OPTIONS]` section for the BDDA to function. The `SEGMENTS` keyword can be used to adjust the accuracy of the water quality simulator. The default value should be adequate unless an EPANET warning is issued stating that the maximum number of segments has been used. The `TOLERANCE` keyword also affects the accuracy of the water quality simulations. Adjusting the tolerance setting in proportion to the injection dosage used for the perturbation analysis will help reduce simulation times. Adjustment of `TOLERANCE` may be needed since its value is interpreted within Epanet as an absolute, not a relative, tolerance.

A.5.2 BDDA input file preparation - MINMASS model

This section details the preparation of the *design* input file (`net2vv.des`) whose sections were described previously. Table A.4 shows the example design input file for the "NET2" network. All existing and potential (new) booster station locations must be identified and listed in the `[INJECTIONS]` section. The Injection types are specified identically for each booster station (`MASS`). The booster scheduling periods, however, have been specified individually. The operation of booster stations `71`, `72`, `291`, `292`, and `261` is modeled using four six-hour boosting periods, with a total schedule cycle length of 24 hours. Injector node `101`, however, uses a schedule that only allows it to operate during the 1st and 3rd six-hour periods. During the 2nd and 4th periods the source is turned off and flow past the booster station is zero. Injections made during these periods would result in large concentration spikes in the system, thus the pattern at node `101` has been specified to avoid this. If the schedule for node `101` were not so modified, the BDDA would still function and return the correct results (i.e. the booster at `101` would be turned off during the 2nd and 4th periods), but such a formulation would be relatively inefficient.

Table A.4: Example BDDA design input file for "NET2"

```
[INJECTIONS]
EXISTING    101
POTENTIAL   71
POTENTIAL   72
POTENTIAL   291
POTENTIAL   292
POTENTIAL   261
TYPE        ALL         MASS
TIME        ALL         24    1 6, 7 12, 13 18, 19 24
TIME        101         24    1 6, 13 18
[CONSTRAINTS]
CONC        ALL         L     0.2
IGNORE      1
IGNORE      26
CONC        101         L     0.0
CONC        101         U     4.0
CONC        71          U     4.0
CONC        72          U     4.0
CONC        291         U     4.0
CONC        292         U     4.0
CONC        261         U     4.0
SELECT      2
[OPTIONS]
```

Table A.4: Example BDDA design input file for "NET2"

```
FORMULATIONS              MINMASS
INFLUENCE                 SAVE  net2vv.inf
MPS          net2vv.mps
[END]
```

Global upper and lower constraint syntax is used to set bounds for all nodes in the [CONSTRAINTS] section. All nodes in the system including tanks, reservoirs, and booster locations are included when the keyword ALL is specified. Once upper and lower concentration limits have been set for all nodes, they can be fine tuned by resetting individual nodes, or using the keyword IGNORE. The constraints written for the tank are ignored and the lower limit specified at node 101 is reset to 0.00 mg/L. This way a minimum disinfectant level is not required at the storage tank or at booster location 101 during periods when no flow is going past the station.

The MINMASS formulation has been selected in the [OPTIONS] section. The formulation will be written in the industry–standard MPS format and the influence file will be saved for later reuse. The influence coefficient file can be reused to quickly generate different optimization problems, however, injector locations and injection scheduling can not be changed, nor can other Epanet settings that would affect the simulated water quality. Constraints can be changed, however, to investigate the effect of different minimum chlorine residual requirements (for example).

A.5.3 Running the BDDA Application

Command line format. The BDDA executable runs from the DOS prompt. Assuming that the current working directory contains the BDDA executable and input files (as well as the DLLs epanet.dll and cplex.dll, if these are not on the machine's DLL search path), the following statement begins execution of the BDDA:

```
bdda.exe bdda.des bdda.rpt epafile.inp epafile.rpt
```

where bdda.exe is the name of the BDDA executable file, bdda.des is the BDDA design input file, bdda.rpt is the BDDA report file, epafile.inp is the Epanet input file, and epafile.rpt is the EPANET report file. Remember that the names of the MPS format output and the impact coefficient binary files are specified in bdda.des.

Solving an LP or MILP. Serious user's are recommended to purchase a CPLEX site license if one is not available. If you have a CPLEX site license you will be able to produce optimal solutions to the BDDA design problems automatically. The optimization problem can be exported to other LP and MILP solver packages using the MPS format file option.

Refer to the instructions for your specific solver package about solving `MPS` format LP and MILP problems.

Interpreting the BDDA Scheduling Solution. Depending on the optimization software used, the solution to the LP or MILP will vary in format. In all cases the part of the solution which contains the scheduling information for each booster station will be in the "COLUMNS" section of the output. In Table A.5 the columns section of the solution output for the example problem is shown. One entry appears for each `XiPj`, where `i` is the booster station location index and `j` is the booster scheduling period index. This scheduling information can be used to modify the Epanet input file so the optimal schedule can be simulated and analyzed further.

Table A.5: Typical format of optimal results ("NET2" example)

```
SECTION 2 - COLUMNS
NUMBER  .....COLUMN.....  AT  ...ACTIVITY...
979     X36P1             LL  0
980     X36P2             LL  0
981     X36P3             BS  0
982     X36P4             LL  0
983     X37P1             BS  785.9914
984     X37P2             BS  81.08658
985     X37P3             BS  197.1814
986     X37P4             LL  0
987     X38P1             BS  0
988     X38P2             LL  0
989     X38P3             BS  0
990     X38P4             LL  0
991     X39P1             BS  0.6109791
992     X39P2             BS  0.2404409
993     X39P3             LL  0
994     X39P4             BS  0.5091436
995     X40P1             BS  215.9495
996     X40P2             BS  394.0498
997     D36               EQ  0
998     D37               EQ  1
999     D38               EQ  0
1000    D39               EQ  1
1001    D40               EQ  1
```

Simulating the Optimal Solution. The columns section of the optimal solution represents the optimal location and scheduling data. This data must be reformatted so that it can be simulated in Epanet. Table A.6 shows the scheduling data from Table A.5 formatted for simulation within a modified Epanet input file; only the `[SOURCES]` and `[PATTERNS]` sections need to be modified, and all other sections are as shown in Table A.3. Injector locations `D40`, `D39`, and `D37` correspond to nodes `101`, `72`, and `292`, respectively. The `[SOURCES]` section of the Epanet input file has been modified to specify these locations as sources. The injection dosages `X40P1` and `P2`, `X39P1` to `P4`, and `X37P1` to `P4` are used to set up the operating schedules for the injectors in the `[PATTERNS]` data. This modified Epanet input file can now be run to simulate and assess performance of the booster system design, and to explore other issues such as design performance under other operating conditions.

Table A.6: Changes to "NET2" example Epanet input file to simulate optimal booster design

```
[SOURCES]
;----------------------------
;           Chlorine    Source
;  ID       mg/L        Pattern
;----------------------------
   101      1.00        3           MASS
   72       1.00        4           MASS
   292      1.00        5           MASS
[PATTERNS]
;Demand pattern
1  1.19      0.97      0.90      0.90      0.82      1.12
1  1.21      0.60      0.60      1.27      2.39      0.90
1  0.85      0.61      1.36      0.54      0.24      0.71
1  0.30      0.60      1.19      1.49      1.12      1.16
;Pump flow pattern
2 0.96 0.96 0.96 0.96 0.96 0.96 0 0 0 0 0 0
2 0.80 1.00 1.00 1.00 1.00 0.15 0 0 0 0 0 0
;Chlorine source pattern node 101
3  215.9495  215.9495  215.9495  215.9495  215.9495  215.9495
3  0         0         0         0         0         0
3  394.0498  394.0498  394.0498  394.0498  394.0498  394.0498
3  0         0         0         0         0         0
;Chlorine source pattern node 72
4  785.9914  785.9914  785.9914  785.9914  785.9914  785.9914
4  81.08658  81.08658  81.08658  81.08658  81.08658  81.08658
4  197.1814  197.1814  197.1814  197.1814  197.1814  197.1814
4  0         0         0         0         0         0
;Chlorine source pattern node 292
5  0.6109791 0.6109791 0.6109791 0.6109791 0.6109791 0.6109791
5  0.2404409 0.2404409 0.2404409 0.2404409 0.2404409 0.2404409
```

Table A.6: Changes to "NET2" example Epanet input file to simulate optimal booster design

```
5  0          0          0          0          0          0
5  0.5091436  0.5091436  0.5091436  0.5091436  0.5091436  0.5091436
```

If CPLEX is used, the optimal solution can be automatically formatted for a modified Epanet input file for simulation, using the EPANET keyword in the [OPTIONS] section of the BDDA input file. The EPANET keyword can also be used without CPLEX to generate a cross reference list that is useful for interpreting the optimization results, if they are obtained using another optimization package (via the MPS option).

A.6 BDDA Warning and Error Codes

Table A.7: BDDA Warning Messages

Warning Code	Warning Message Description
201	BDDA input file formatting error
202	Invalid value for SELECT parameter
203	No injector locations have been selected
204	Unable to get Epanet index value for Node ID
205	Pattern start or stop time greater than pattern duration
206	Upper and lower concentration constraints reversed
208	Unable to scan parameter value

Table A.8: BDDA Error Messages

Error Code	Error Message Description
510	Unable to open BDDA input file
511	Unable to open BDDA report file
512	Unable to open MPS format output file
513	Failure to Writestring to file
514	Unidentified injection type
515	Unable to open Epanet temp file
517	Unable to write Epanet temp file
520	qnode vector length in error
610	Unable to open influence file for wrirting
611	Unable to open influence file for reading
612	Influence file write error
613	Influence file read error

Table A.8: BDDA Error Messages

Error Code	Error Message Description
615	Invalid influence file
616	Unable to compute influence file position
617	Influence file read returned wrong column
620	Influence file column length in error
700	Unable to open Epanet input file for copying
705	Unable to open optimized Epanet input file for pasting
900	Memory allocation error
910	Memory reallocation error
**	Abnormal termination in BDDA, check Epanet or CPLEX documentation for details

Appendix B

Booster Disinfection Utility Surveys

B.1 Postcard Survey (Survey 1)

Dear Utility,

This is a brief survey for a booster disinfection project sponsored by the American Water Works Association Research Foundation. Your participation in this survey is very important and will only take a few minutes of your time.

As you may be aware, booster disinfection – the application of disinfectant at points located throughout the distribution system – has a number of benefits. It has the potential to satisfy disinfectant residual requirements and provide more uniform disinfectant coverage, while reducing the total amount of disinfectant required.

A reduction in the amount of disinfectant required may cause a decrease in disinfection by-product formation; thus, an improvement in water quality. Also, uniform disinfectant coverage could help prevent water quality problems related to bacterial regrowth in distribution systems.

With your help, our work will quantify these benefits of booster disinfection and develop design methods for operating and locating booster stations. Only through your participation can this work be a success.

Thank you very much for your participation!

Sincerely,

Dr. James Uber
Project Primary Investigator
University of Cincinnati

Survey Instructions

Please fill out the contact information and answer the questions as completely as possible. When you are finished, detach the answer card and return to us. No postage is necessary. Thank you!

Survey (Please print.)

1. How many consumers does your utility serve?
   ☐ 10,000 or less ☐ 10,001 - 50,000 ☐ 50,001 - 100,000
   ☐ 100,001 - 500,000 ☐ more than 500,000

2. How many different sources of water feed into your distribution system?
   ____

3. For each different source of water feeding your distribution system, identify the type, the primary disinfectant and the percent usage (of total demand) over the year.

Source type	Primary disinfectant	Percent usage by month (Jan Feb Mar Apr May June July Aug Sept Oct Nov Dec)

Example for filling out question three:

A utility uses chlorinated river water as a primary source and augments it with chloraminated groundwater during the high-demand summer season.

Source type	Primary disinfectant	Percent usage by month (Jan Feb Mar Apr May June July Aug Sept Oct Nov Dec)
River	*Chlorine*	100% ⟶ 85% ⟶ 100% ⟶
Groundwater	*Chlorine*	15% ⟶

4. If you use multiple groundwater sources, do you consider the water quality (e.g., pH, alkalinity, temperature, DO, TOC) to be similar at all the wells?
   ☐ yes ☐ no ☐ not applicable

5. Do you currently booster disinfect *previously disinfected water* at any location in the distribution system (i.e., after leaving the treatment plant)?
   ☐ yes ☐ no

6. If you answered "yes" to question five, what was the primary reason leading to the installation of these booster disinfection stations?
   ☐ historic residual low point ☐ taste and odor control
   ☐ discoloration ☐ biological regrowth prevention
   ☐ disinfection by-product control ☐ nitrification
   ☐ other____________________

7. Are you interested in methods of *disinfecting treated water in the distribution system*, in addition to disinfection at the treatment plant?
   ☐ yes ☐ no ☐ don't know

Contact Information

Name ____________________ Title ____________________

Utility Name ____________________

Street Address ____________________

City, State, ZIP ____________________

Phone ____________________ Fax ____________________

E-mail ____________________

UCPub8581

B.2 Booster Disinfection Survey (Survey 2)

Current Practices in Booster Disinfection

The University of Cincinnati
Department of Civil & Environmental Engineering

An AWWARF Sponsored Survey of Selected Water Utilities

Who Should Fill Out The Survey

Please direct this survey to a person with knowledge of water quality and distribution system design and operation. This person should have general knowledge of disinfection practices and of booster station design and operation.

Survey Directions

General

The survey includes three types of questions. One requires written answers — please note that specific units are sometimes requested. Another requires you to check one from a set of possible answers. *Please check only one answer unless asked to do otherwise.* Finally, some questions ask you for a scaled response — the scale is always from 1 through 5, and you are to circle the most appropriate number.

Throughout the survey we make reference to chlorine as a disinfectant. It may be inaccurate, but we have used hypochlorous acid and chlorine interchangeably. Please keep this in mind when reading the questions.

Part A

In part A we ask a few brief questions to characterize your utility's treatment practices, distribution system, and safety issues concerning booster station operation.

Part B

In part B we ask more detailed questions about the design and operation of *up to* two (2) specific booster disinfection stations. If your distribution system includes more than two booster stations, then please select just two to describe; to maximize the value of the information, select two booster stations that are different from each other in terms of location, design, or operation (if possible).

Please answer as many questions as possible, skipping those which do not apply. A partially completed survey is still useful to us. It should be stressed that all responses will be held in confidence. Thank you in advance for your participation.

Current Practices in Booster Disinfection

Part A

Contact Information

Name:____________
Title:____________
Utility Name:____________
Street Address:____________
City, State and Zip:____________
Phone:____________
Fax:____________
Email:____________
May we contact you again if needed? ❑Yes ❑No

Treatment

1. What is the annual average finished water production rate, in MGD?____________
2. Does your treatment plant regularly apply a disinfectant before sedementation (pre-oxidant) and/or after sedementation (post-oxidant)?
   ❑ Post-oxidant alone
   ❑ Pre and Post-oxidant
   ❑ Pre-oxidant application varies seasonally

Note: If you apply a pre-oxidant answer questions 3-5, else go to question 6.

3. Which pre-oxidant is applied the majority of the time?
   ❑ Chlorine gas ❑ Chloramine ❑ Ozone
   ❑ Hypochlorite solution
   ❑ Potassium permanganate
   ❑ Other____________
4. What is the single most important factor in determining the pre-oxidant dose used?

Question 4 continued:
❑ Taste and odor control
❑ DBP formation ❑ C*T Requirements
❑ Residual maintenance
❑ Enhanced coagulation
❑ Other____________

5. Which post-oxidant is applied the majority of the time?
   ❑ Chlorine gas ❑ Preformed chloramines
   ❑ Hypochlorite solution
   ❑ Chlorine gas and ammonia
   ❑ Hypochlorite and ammonia
   ❑ Other____________
6. What is the single most important factor governing post-oxidant dose?
   ❑ C*T Requirement ❑ DBP Formation
   ❑ Residual maintenance in dist. system
   ❑ Other____________
7. For post-oxidant applications, please estimate the total amount applied annually and the unit cost (indicate units).
   Total applied____________
   Unit Cost____________
8. How is contact time, T_{10}, determined for the process train where disinfection is credited?
   ❑ Tracer test performed on each unit process and resulting T_{10}'s summed
   ❑ SWTR "rule of thumb" applied for each unit process and resulting T_{10}'s summed
   ❑ Tracer test performed for entire process train resulting in a single T_{10} value
   ❑ Other____________
9. What are the *actual* conditions in the disinfectant contact chamber used to calculate $C*T_{10}$ achieved before the first consumer? Please respond with average, maximum, and minimum C*T conditions on the table below.

Note: Table for Ques. 9	Average	Max	Min
Flow Rate (MGD)			
Water Temp (°C)			
Water pH			
T_{10} Achieved (min)			
Disinfectant Conc (mg/l)			
C*T Achieved(mg*min/L)			
C*T Required (mg*min/L)			

Current Practices in Booster Disinfection

Part A - Continued

Distribution System

1. What is the total population served?
   ❑ <10,000 ❑ 10,000-50,000
   ❑ 50,001-100,000
   ❑ 100,001-500,000 ❑ >500,000
2. What is the total number of separate booster stations that are installed and operational, the number operated continuously, and the number operated seasonaly?
   Total__________
   Continuously__________
   Seasonaly__________
3. What is the number of finished water storage tanks and reservoirs?__________
4. What is the average finished water volume stored in tanks and reservoirs, in millions of gallons?__________
5. What is the total length of distribution system piping, in miles?__________
6. What is the *maximum* travel time from a point of treatment to a consumer, in days?
   __________ ❑ Don't know
7. Characterize the average disinfectant demand in the *bulk* finished water:
   Low Demand 1 2 3 4 5 High Demand
   ❑ Don't know
8. Characterize the average disinfectant demand associated with *pipe wall* reactions in the distribution system (wall demand):
   Low Demand 1 2 3 4 5 High Demand
   ❑ Don't know
9. Characterize the average age of distribution system pipes:
   ❑ < 20 years ❑ 20 to 40 years
   ❑ 40 to 60 years ❑ 60 to 80 years
   ❑ > 80 years ❑ Don't know
10. What particular pipe materials are associated with the highest pipe wall demands found in your system?
    ❑ Concrete/Concrete Lined
    ❑ Asbestos cement/Asbestos cement Lined
    ❑ Polyethylene/Polyethylene Lined
    ❑ Ductile/Cast Iron ❑ Don't know
    ❑ Other__________
11. What percentage of your distribution system uses these "high demand" material?
    ❑ < 5% ❑ 5 to 20% ❑ 20 to 40%
    ❑ 40 to 60% ❑ 60 to 80% ❑ >80%
    ❑ Don't know
12. Does your distribution system currently have any of the following problems (check all that apply)?
    ❑ Biological regrowth ❑ Taste and odor
    ❑ Stagnant water ❑ Discoloration
    ❑ Low disinfectant residuals
    ❑ High disinfection by-product formation
    ❑ Excess, or ❑ Inadequate storage capacity
    ❑ Low pressure ❑ Nitrification
    ❑ Other__________

Safety

1. What types of safety equipment are installed at the booster station (check all that apply)?
   ❑ Spill kits ❑ Gas scrubbers
   ❑ Respirators ❑ Personal protective gear
   ❑ Other__________
2. Describe briefly any special regulatory requirements that apply to the operation of the booster station such as RCRA, OSHA, RMP, State Regulations, or Public Health codes.
   ____
   ____
   ____
   ____
3. Have you found that safety issues have made the operation of the booster stations cost prohibitive or restrictive?
   ❑ Yes ❑ No ❑ Don't know
4. If you answered "yes" to question 3, briefly describe what safety issues were costly or overly restrictive.
   ____
   ____
   ____
   ____
5. Have you ever had an OSHA reportable accident associated with the operation of your booster stations?
   ❑ Yes ❑ No ❑ Don't know
6. Have you ever had a RCRA reportable spill or leak associated with the operation of your booster station?
   ❑ Yes ❑ No ❑ Don't know

Part B - Booster Station #1

General

1. How many years has this booster station been in operation?__________
2. This booster station was installed to control:
   ❑ Low disinfectant residual
   ❑ Nitrification
   ❑ Disinfectant by-product formation
   ❑ Taste and odor ❑ Discoloration
   ❑ Biological regrowth
   ❑ Other__________
3. Indicate your overall satisfaction with this booster station's ability to achieve the treatment objective specified in question 2:
   Not Satisfied 1 2 3 4 5 Very Satisfied
4. Describe briefly any other particular benefits attributed to installation of this booster station.
5. Describe briefly any particular problems attributed to installation of this booster station.
6. How would you rate the reliability of equipment used in this booster station?
   Unreliable 1 2 3 4 5 Very reliable
7. Describe briefly the types and frequency of routine maintenance required to operate the booster station.
8. Describe briefly any startup problems associated with the installation of your booster stations.
9. Describe any types of non-routine maintenance that your booster station has required.

Location and Design

1. During an average day, please characterize the water flow rate at this booster location:
   Constant 1 2 3 4 5 Highly Variable
2. During an average day, please characterize the disinfectant concentration at this booster location (*prior* to boosting):
   Constant 1 2 3 4 5 Highly Variable
3. Does the water routinely reverse flow direction at this booster location?
   ❑ Yes ❑ No ❑ Don't know
4. Does this booster station serve a distribution system zone that is practically isolated from the influence of any other booster station?
   ❑ Yes ❑ No ❑ Don't know
   ❑ No other operable booster stations
5. What information was used in selecting this booster station location?
   ❑ Proximity to existing pump stations
   ❑ Proximity to storage tanks, or reservoirs
   ❑ Operating experience
   ❑ Proximity to areas with low residuals
   ❑ Distribution system hydraulic and water quality simulation model results
   ❑ Other__________
6. If this booster station is located at a finished water storage tank or reservoir, where is it located?
   ❑ A common inlet/outlet line
   ❑ A separate inlet line
   ❑ A separate outlet line
   ❑ A dedicated recirculation line
   ❑ Other__________
7. Is the booster station located ❑ above or ❑ below grade?
8. What is the design life of this booster station, in years? __________
9. Did you purchase a prepackaged booster station system?
   ❑ Yes ❑ No ❑ Don't know

10. If you answered no to question 9, then who designed the booster station?
❑ In-house engineering or operations personnel
❑ A consulting engineer
❑ Vendor engineering support services
❑ Other______________________

11. What is the physical size of this booster station site in feet?
Width ________
Length________
Height________

Disinfectant Type & Dose

1. What is the type of disinfectant is applied at the booster station?
❑ Chlorine gas ❑ Hypochlorite solution
❑ Pre-formed chloramines
❑ Separate chlorine and ammonia
❑ Other______________________

2. How are disinfectants stored?
❑ Bulk dry feed chemicals
❑ Compressed or liquified gas
❑ Bulk chemical solutions
❑ On-site generation
❑ Other______________________

3. Enter the average water flow rate leaving the booster station, in mgd:______________

4. This question pertains to the booster disinfectant *feed solution(s)*. Enter the percent, by weight, of hypochlorite and aqueous ammonia in the feed solution(s), plus the feed solution usage rate(s) in lbs/week:
wt% HOCl_____ lbs/week___________
wt% NH_3_____ lbs/week___________

5. These next two questions pertain to the disinfectant residual before and after boosting. Enter the average residual disinfectant and ammonia concentrations as they *enter* the booster station: (enter "unknown" if these data are not available):
1. Free available chlorine (mg/l)________
2. Combined avail. chlorine (mg/l)_______
3. NH_3 (mg/l)_________________

6. Enter the average residual disinfectant and ammonia concentrations as they *leave* the booster station, after boosting: (enter "unknown" if these data are not available):
1. Free available chlorine (mg/l)_________
2. Combined chlorine (mg/l)___________
3. NH_3 (mg/l)__________________

7. If your post-oxidant is chloramines do you boost through breakpoint chlorination to produce a free chlorine residual?
❑ Yes ❑ No ❑ Not applicable

8. If you answered no to question 7, do you add chlorine to?
❑ Increase Chlorine to Nitrogen ratio
❑ Maintain Chlorine to Nitrogen ratio
❑ Recombine free Nitrogen
❑ Other______________________

Operation

Questions 1-5 pertain to water quality sampling that is routinely performed for monitoring the performance of this booster station and making operational changes. *Please do not include monitoring performed for regulatory or other purposes.* (In each case, check all that apply, or "Not applicable" if no such samples are collected).

1. Enter the location and frequency of samples analyzed for *free available chlorine*:
❑ Not applicable
Location:
❑ Just upstream of injection
❑ Just downstream of injection
❑ In distribution system zone downstream of booster
Frequency:
❑ continuous analyzer ❑ daily ❑ weekly
❑ monthly ❑ >monthly

2. Enter the location and frequency of samples analyzed for *combined chlorine*:
❑ Not applicable
Location:
❑ Just upstream of injection
❑ Just downstream of injection
❑ In distribution system zone downstream of booster
Frequency:
❑ continuous analyzer ❑ daily ❑ weekly
❑ monthly ❑ >monthly

3. Enter the location and frequency of samples analyzed for *ammonia*:
   ❑ Not applicable
   Location:
   ❑ Just upstream of injection
   ❑ Just downstream of injection
   ❑ In distribution system zone downstream of booster
   Frequency:
   ❑ continuous analyzer ❑ daily ❑ weekly
   ❑ monthly ❑ >monthly

4. Enter the location and frequency of samples analyzed for *oxidation reduction potential (ORP)*:
   ❑ Not applicable
   Location:
   ❑ Just upstream of injection
   ❑ Just downstream of injection
   ❑ In distribution system zone downstream of booster
   Frequency:
   ❑ continuous analyzer ❑ daily ❑ weekly
   ❑ monthly ❑ >monthly

5. Enter the location and frequency of other sampled quantities:
   Measurement:______________
   or ❑ Not applicable
   Location:
   ❑ Just upstream of injection
   ❑ Just downstream of injection
   ❑ In distribution system zone downstream of booster
   Frequency:
   ❑ continuous analyzer ❑ daily ❑ weekly
   ❑ monthly ❑ >monthly

6. Is the water flow rate leaving this booster station measured continuously?
   ❑ Yes ❑ No

7. For this booster station, the main *objective* of feed control is:
   ❑ Maintain time-varying "set-point" conc.
   ❑ Maintain constant "set point" conc.
   ❑ Maintain constant feed rate(s)
   ❑ Other____________________

8. The *method* of feed control is:
   ❑ On-site manual control
   ❑ Remote manual control (SCADA)
   ❑ Automatic (continuous) control
   ❑ Other____________________

9. If you selected "automatic control" above, then please answer these next two questions, else go to the following section. The method of automatic control is:
   ❑ Disinfectant conc. feedback control
   ❑ Disinfectant conc. and flow rate feedforward control
   ❑ Flow proportional ("flow pacing") control
   ❑ Other____________________
   ❑ Don't know

10. The following special safety precautions were used when implementing automatic control (check all that apply):
    ❑ Redundant sensors
    ❑ High/Low dose cutoffs
    ❑ Remote alarms ❑ Failsafe features
    ❑ Other____________________
    ❑ Don't know

Cost

1. Please estimate the total design, construction, and equipment costs associated with installing this booster station.
   $____________________

2. Please estimate the annual maintenance and operating costs associated with this booster station, *excluding chemical costs.*
   $____________________

3. Please estimate the unit cost for the disinfectant applied at the booster station and the total amount applied on an annual basis.
   Unit Cost____________(Indicate Units)
   Total applied____________(Indicate Units)

Appendix C

Additional Experimental Results

This appendix contains additional experimental data related to the material in chapter 4, but which were not specifically discussed there. The related figures from chapter 4 are listed in parentheses for each figure, or set of figures.

- (Figure 4.6). Figures C.1 and C.2 show disinfectant concentration for treated Harsha Lake water of the Phase I.a experiments; Cl:DOC of 2.0:1 and 1.5:1.
- (Figure 4.7). Figures C.3 and C.4 show disinfectant demand for treated Harsha Lake water of the Phase I.a experiments; Cl:DOC of 2.0:1 and 1.5:1.
- (Figure 4.9). Figures C.5 and C.6 show disinfectant demand after boosting for treated Harsha Lake water of the Phase I.a experiments; initial Cl:DOC of 2.0:1 and 1.5:1.
- (Figure 4.20). Figures C.7 and C.8 show disinfectant demand for conventionally and conventionally/ozonated treated Harsha Lake water of the Phase II.a experiments; Cl:DOC of 1.5:1.
- (Figure 4.21 and 4.22). Figures C.9 and C.10 show disinfectant concentration for South Norwalk water of the Phase II.a experiments; Cl:DOC of 3:1 and 2:1.
- (Figure 4.23 and 4.24). Figures C.11 and C.12 show disinfectant demand for South Norwalk water of the Phase II.a experiments; Cl:DOC of 3:1 and 2:1.
- (Figures 4.31 and 4.32). Figure C.13 shows disinfectant concentration and THM formation for South Norwalk water of the Phase II.b experiments, under multiple boost doses.
- (Figure 4.33). Figure C.14 shows disinfectant demand for South Norwalk water of the Phase II.b experiments, under multiple boost doses.

- (Figure 4.34). Figures C.15 and C.16 show disinfectant concentration and THM formation for Cincinnati water of the Phase II.b experiments, with single ($B_0 = 0.68M_0$) and multiple boost doses.
- (Figure 4.35). Figure C.17 shows disinfectant demand for Cincinnati water of the Phase II.b experiments, with single ($B_0 = 0.68M_0$) and multiple boost doses.
- (Figures 4.36 and 4.37). Figure C.18 shows disinfectant concentration and THM formation for Midland water of the Phase II.b experiments, with multiple boost doses.
- (Figure 4.38). Figure C.19 shows disinfectant demand for Midland water of the Phase II.b experiments, with multiple boost doses.
- (Figure 4.39). Figure C.20 shows disinfectant demand for Las Vegas water of the Phase II.b experiments (pH = 8.0), with a single boost dose ($B_0 = 0.83M_0$).

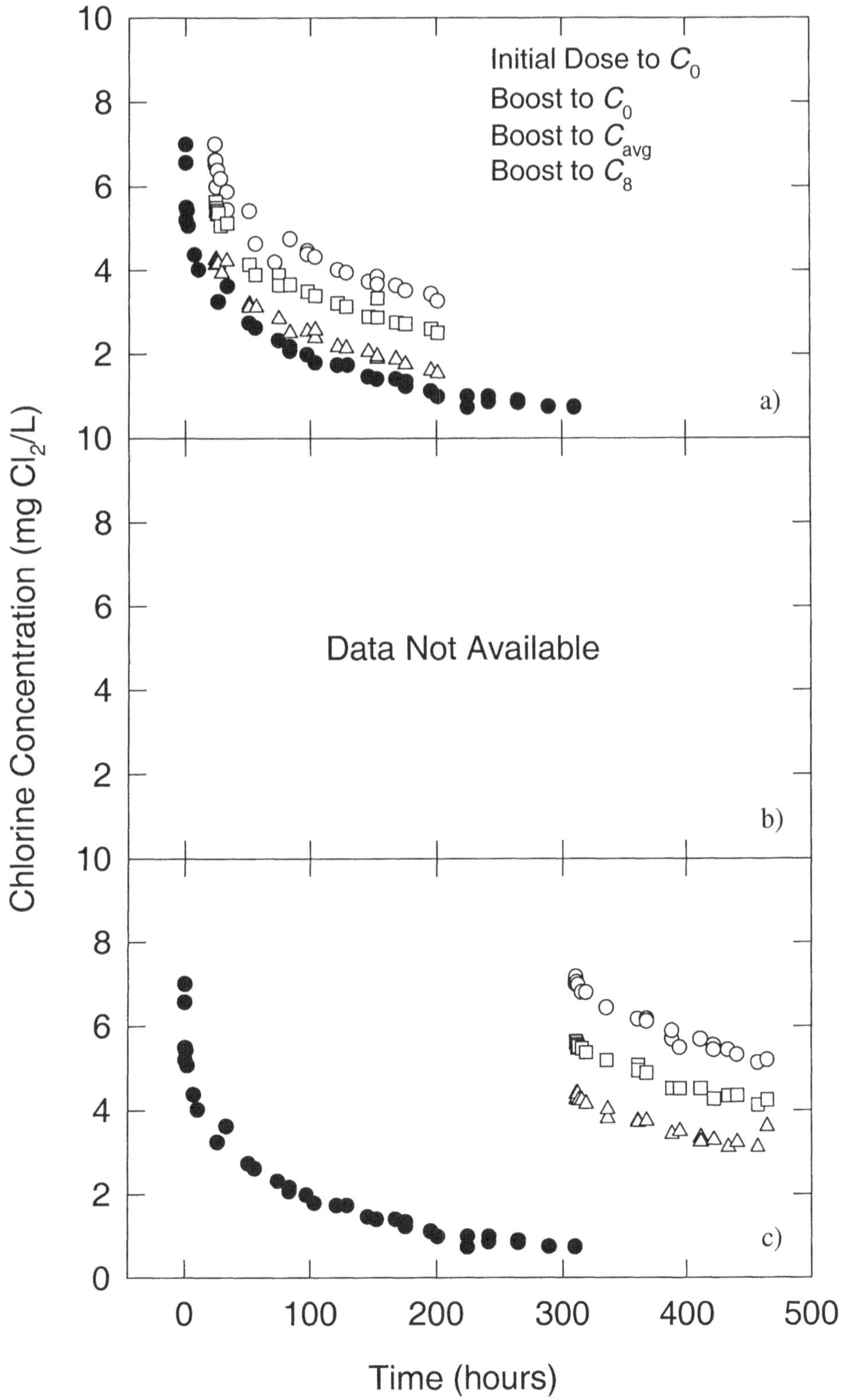

Figure C.1: Chlorine decay data of Phase Ia based on an initial Cl:DOC of 2.0:1 with boost doses at: a) 24 hours, b) 48 hours, and c) when the residual of the initial dose is 0.2 mg/L.

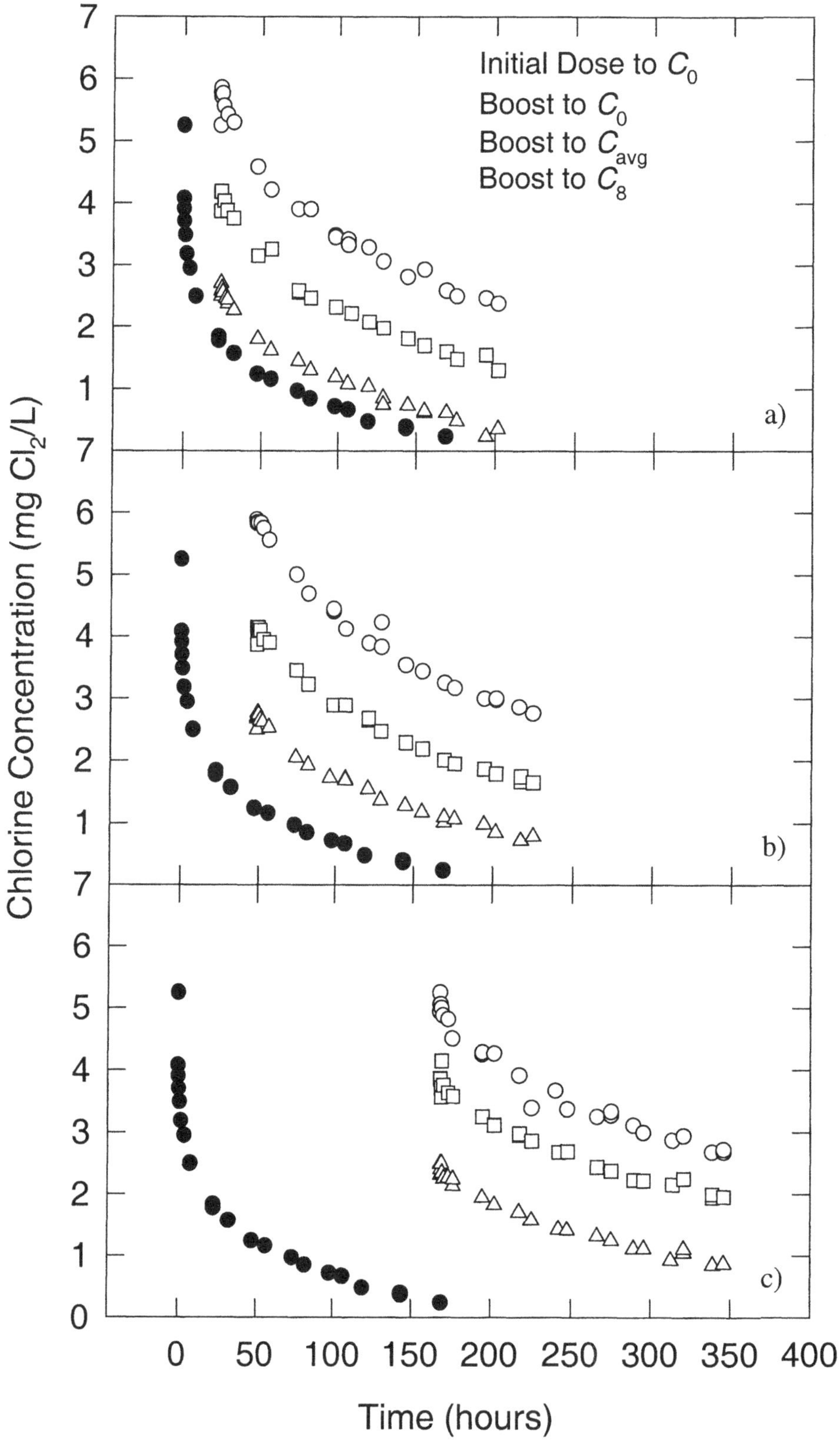

Figure C.2: Chlorine decay data of Phase Ia based on an initial Cl:DOC of 1.5:1 with boost doses at: a) 24 hours, b) 48 hours, and c) when the residual of the initial dose is 0.2 mg/L.

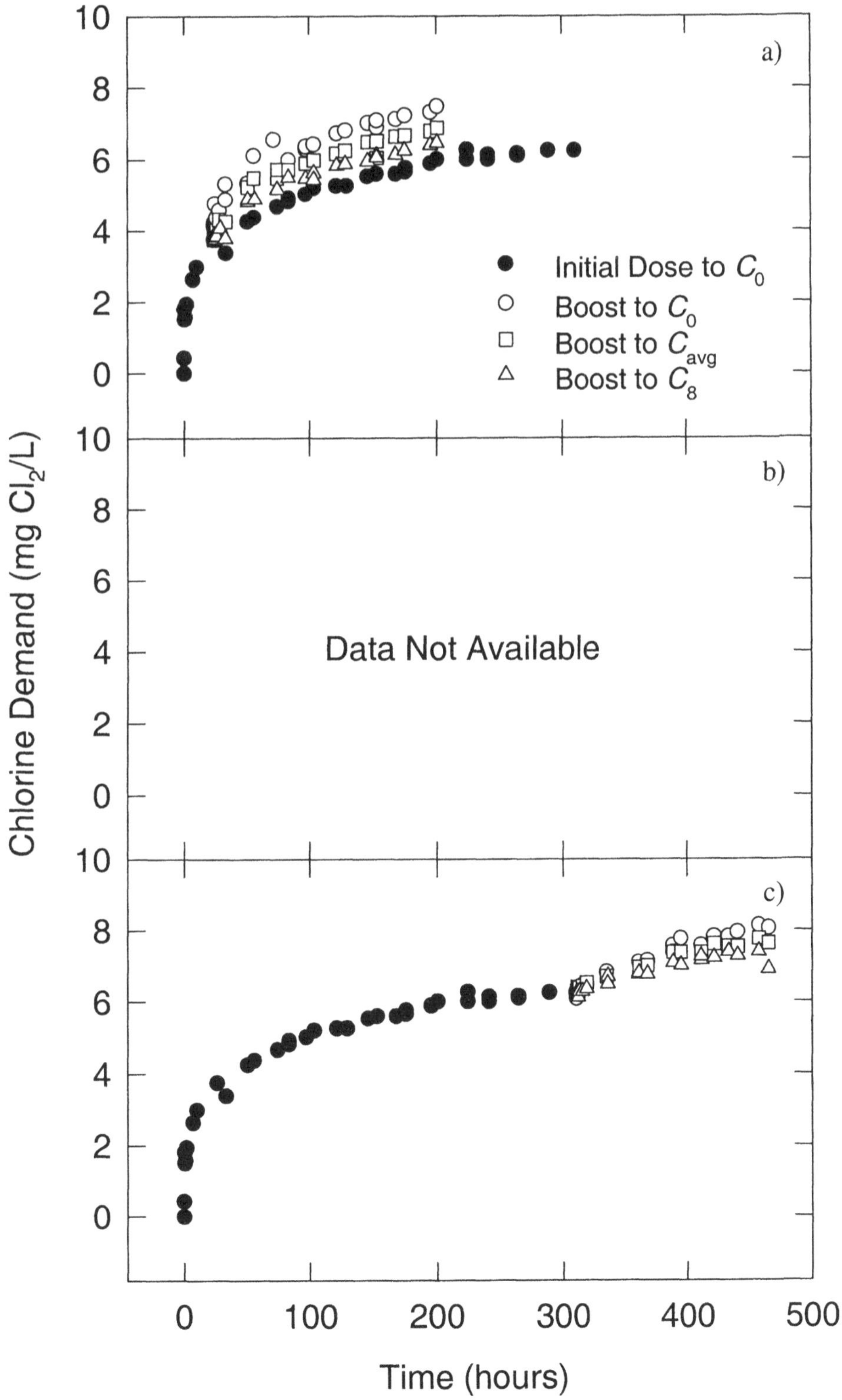

Figure C.3: Chlorine demand data of Phase Ia based on an initial Cl:DOC of 2.0:1 with boost doses at: a) 24 hours, b) 48 hours, and c) when the residual of the initial dose is 0.2 mg/L.

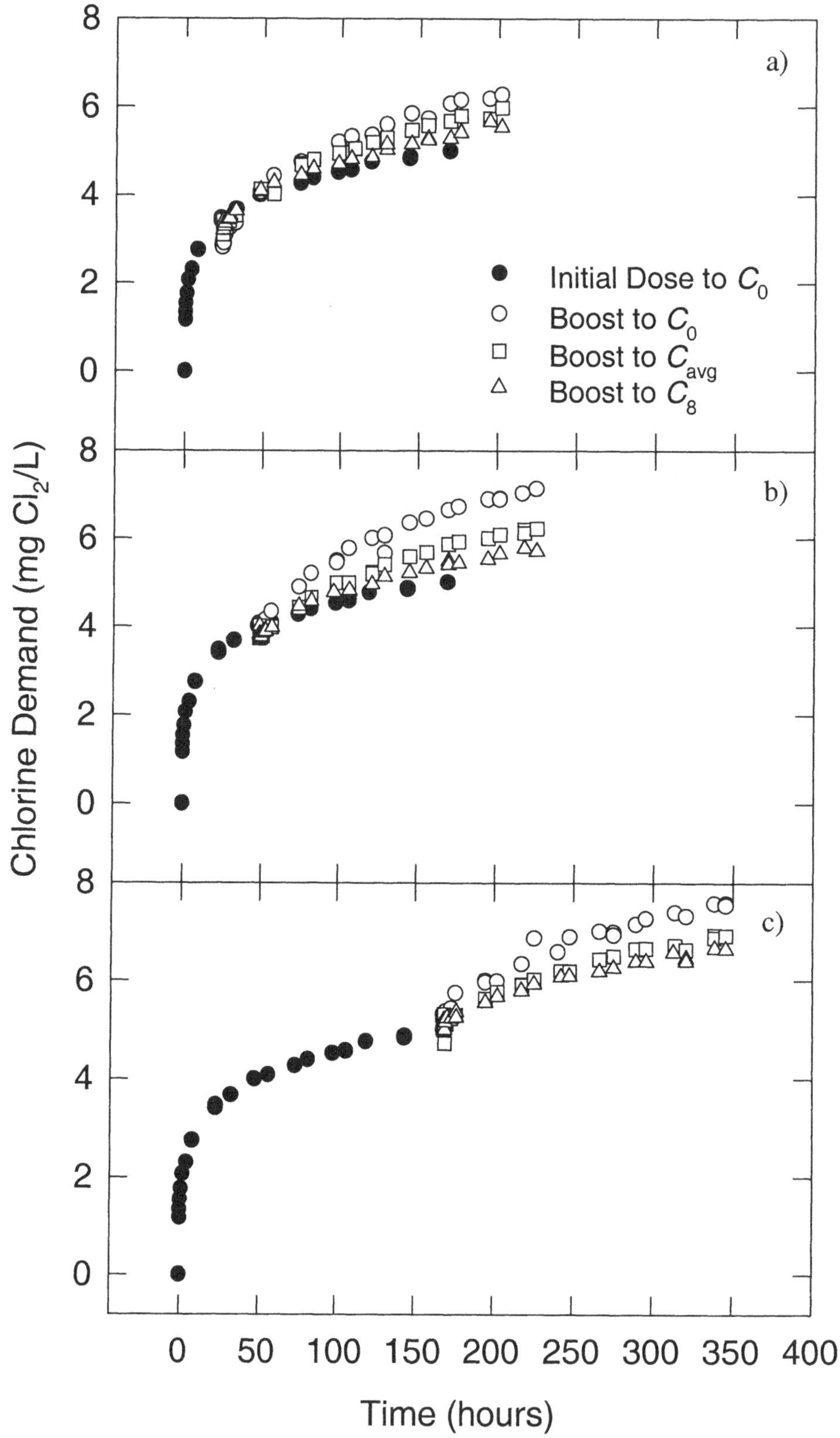

Figure C.4: Chlorine demand data of Phase Ia based on an initial Cl:DOC of 1.5:1 with boost doses at: a) 24 hours, b) 48 hours, and c) when the residual of the initial dose is 0.2 mg/L.

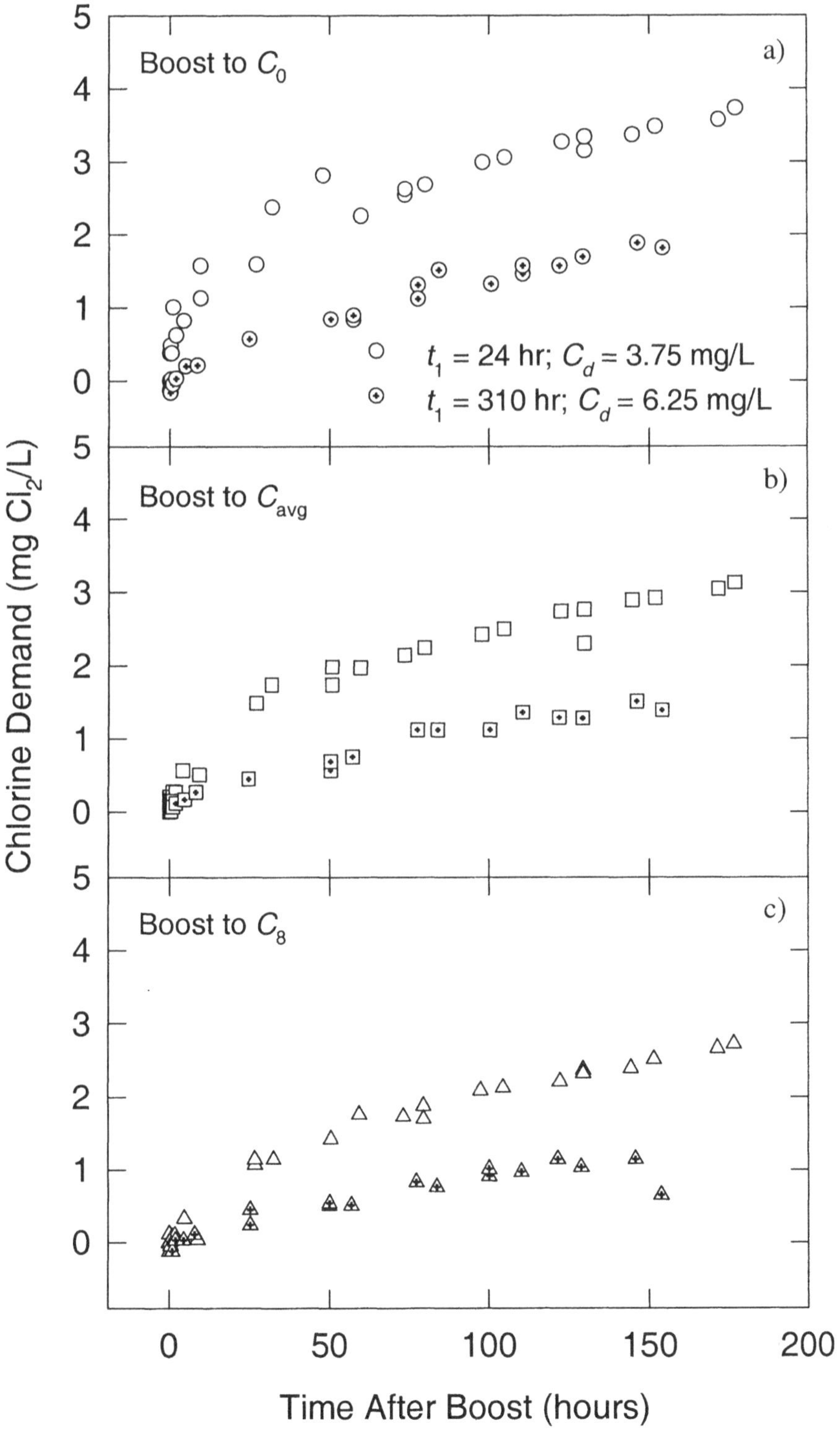

Figure C.5: Chlorine demand data following one of three boost doses of Phase Ia based on a Cl:DOC of 2.0:1.

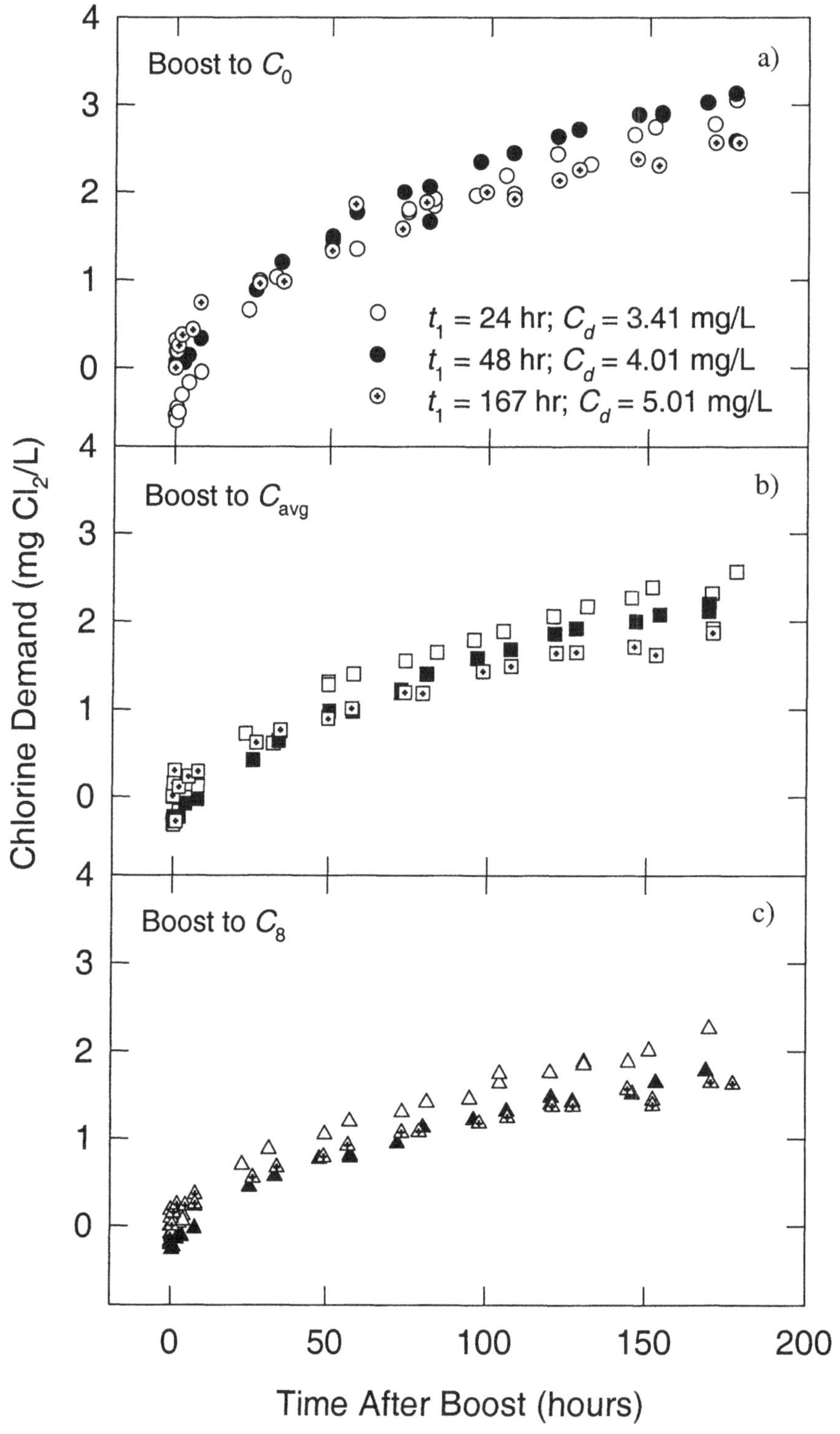

Figure C.6: Chlorine demand data following one of three boost doses of Phase Ia based on a Cl:DOC of 1.5:1.

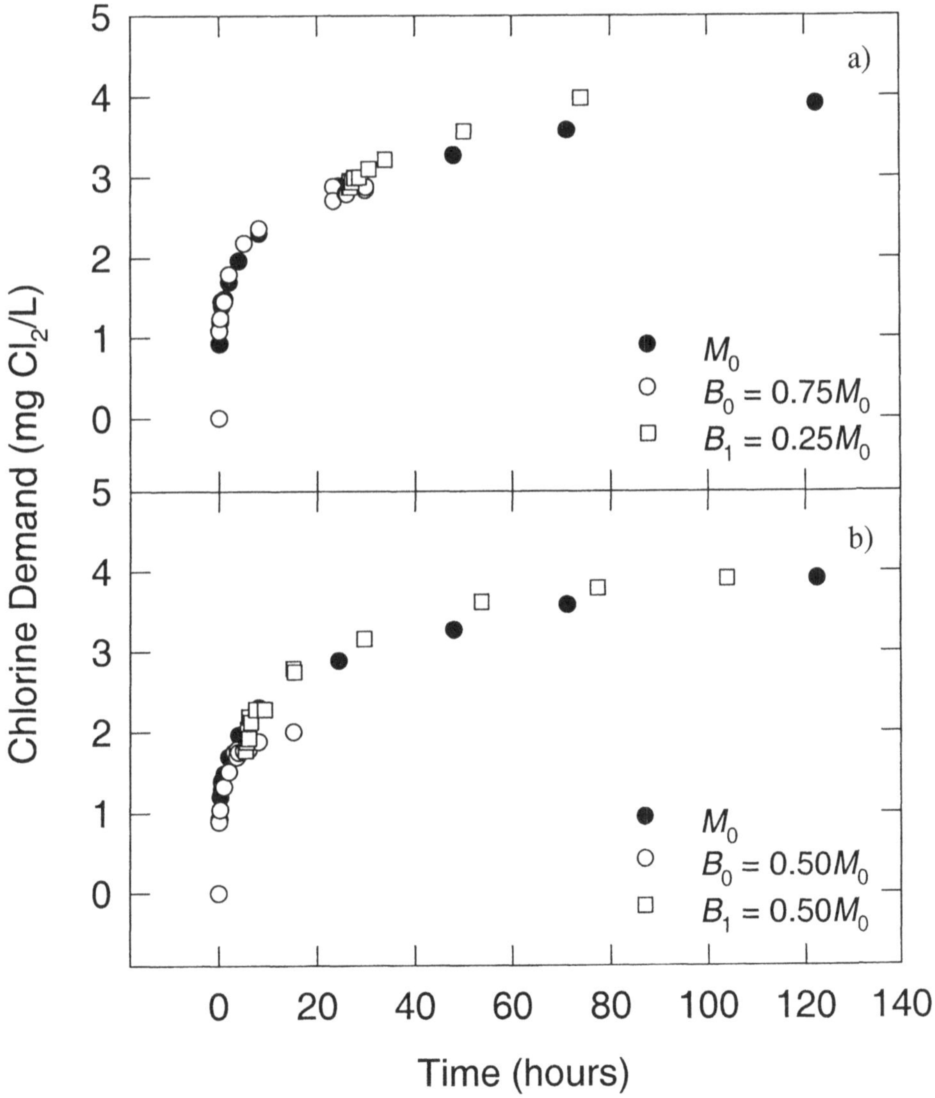

Figure C.7: Chlorine demand data of phase IIa for conventionally treated Harsha Lake water with B_0 determined by a) $0.75M_0$ and b) $0.50M_0$.

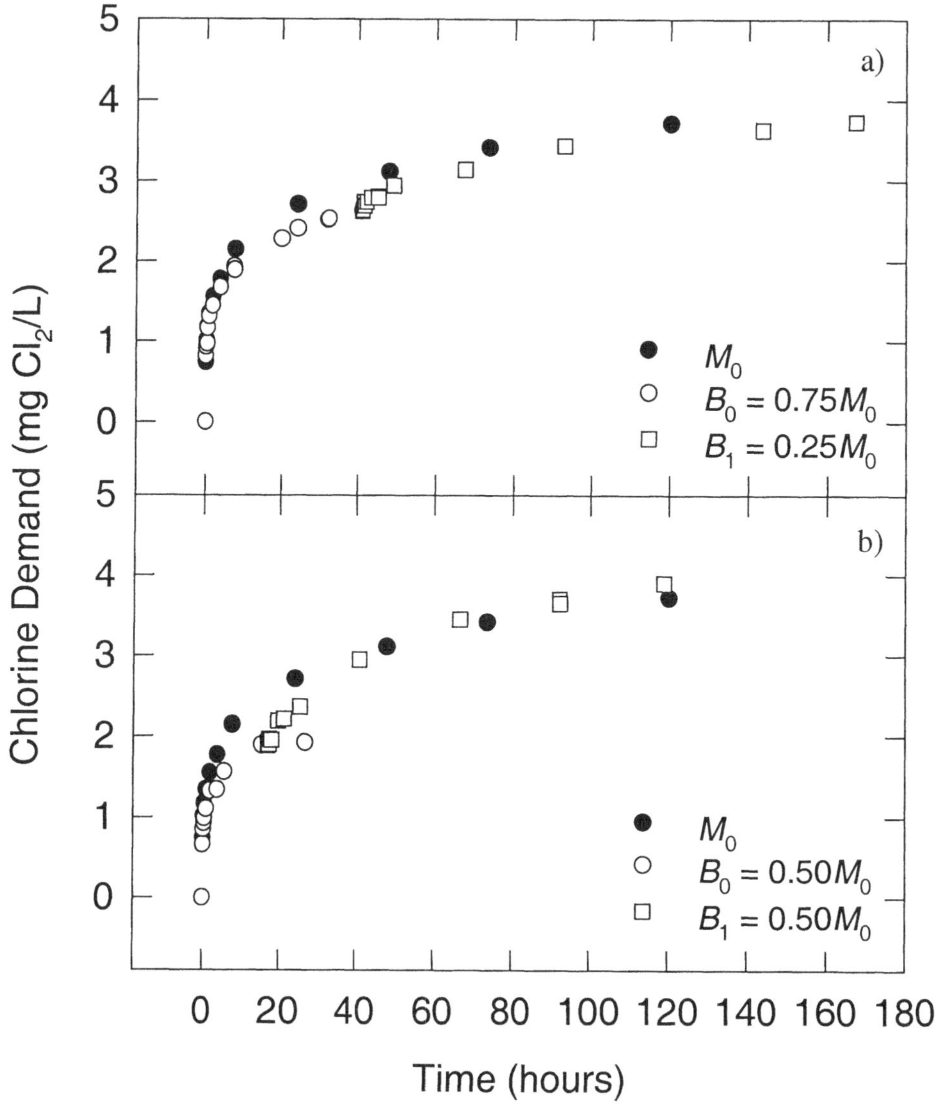

Figure C.8: Chlorine demand data of phase IIa for conventional/ozonated treated Harsha Lake water with B_0 determined by a) $0.75M_0$ and b) $0.50M_0$.

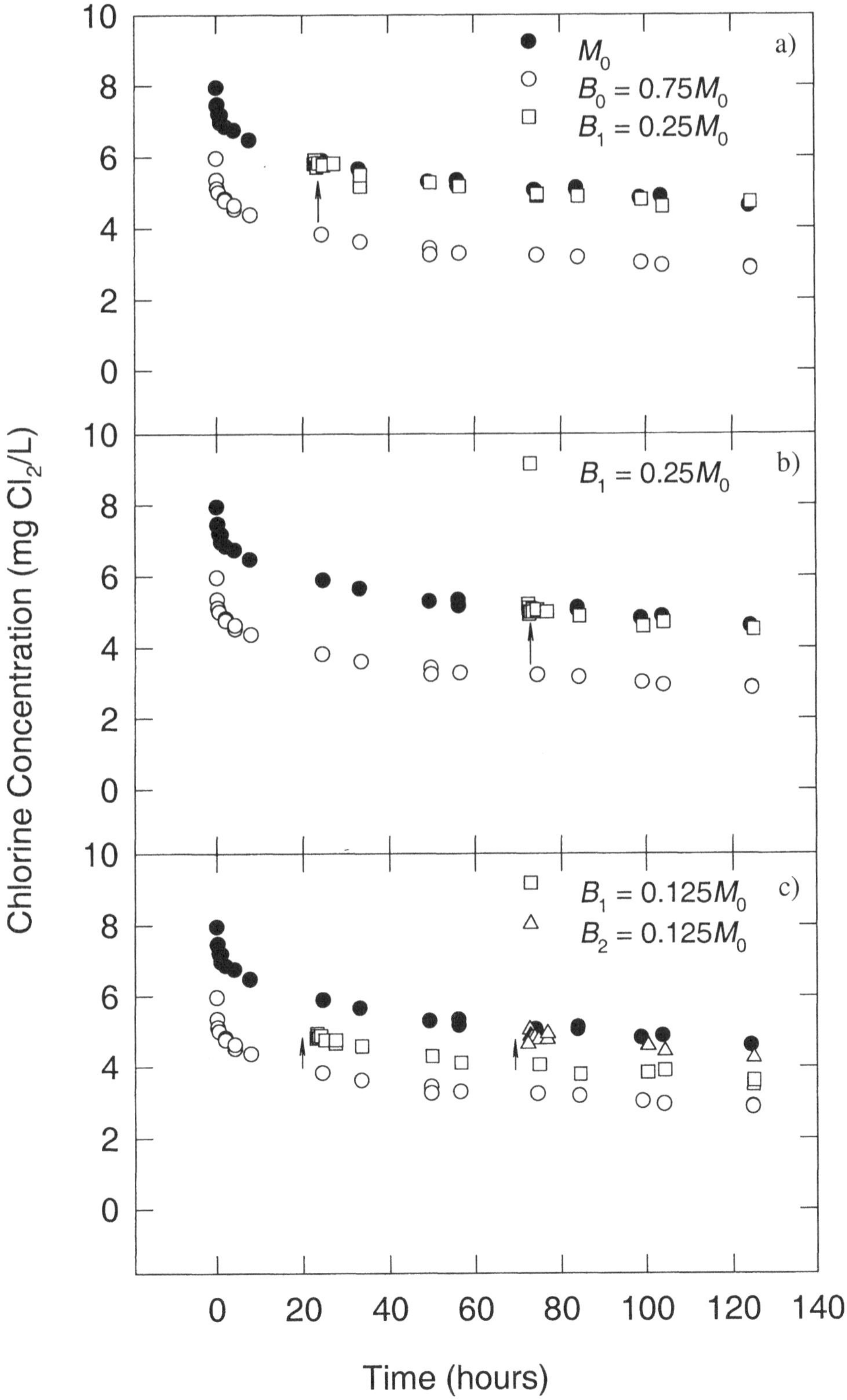

Figure C.9: Chlorine decay data of phase IIa for South Norwalk water based on a Cl:DOC of 3:1 and $B_0 = 0.75M_0$ with boost doses at a) 24 hours, b) 72 hours, and c) 24 and 72 hours.

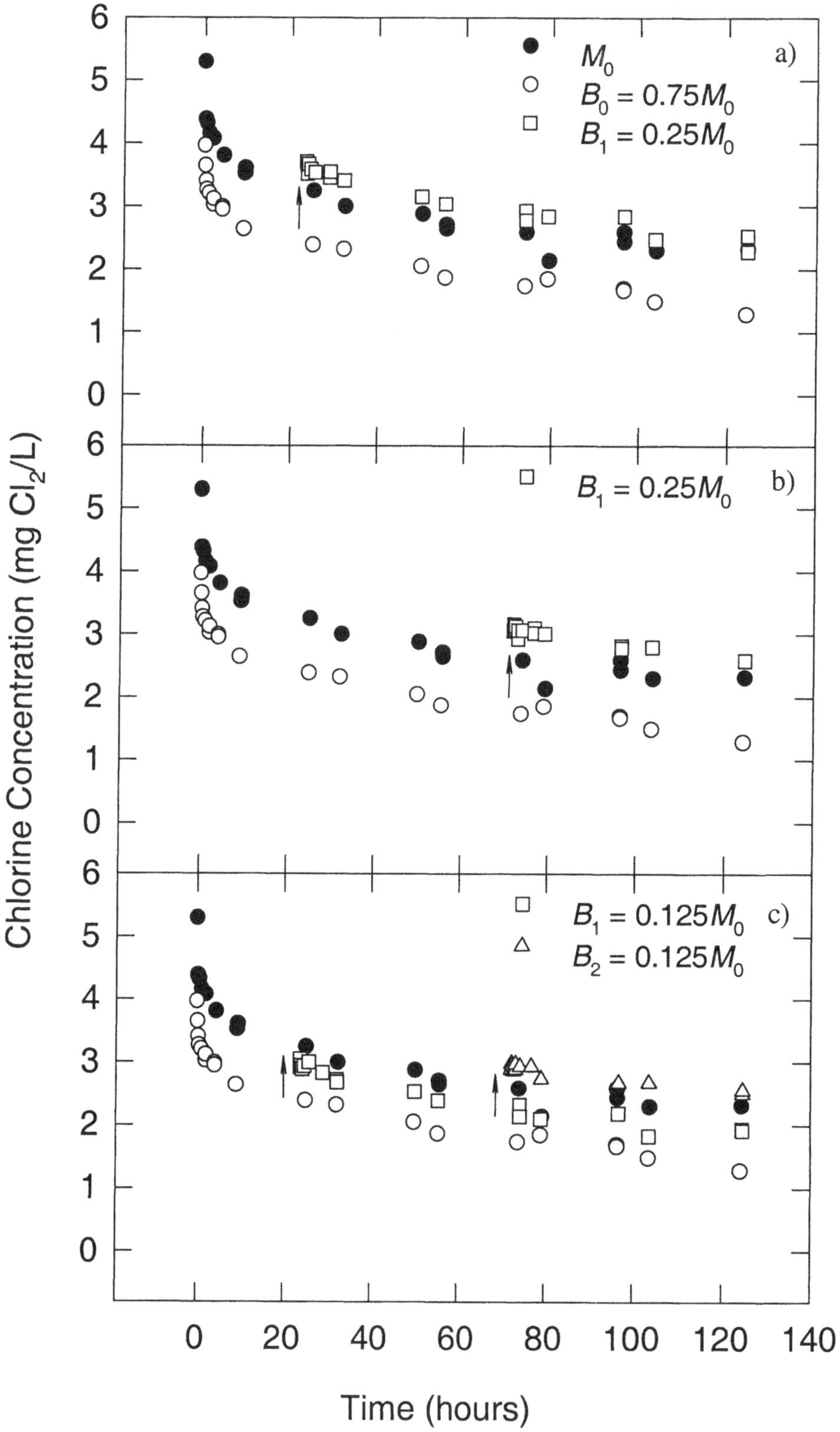

Figure C.10: Chlorine decay data of phase IIa for South Norwalk water based on a Cl:DOC of 2:1 and $B_0 = 0.75M_0$ with boost doses at a) 24 hours, b) 72 hours, and c) 24 and 72 hours.

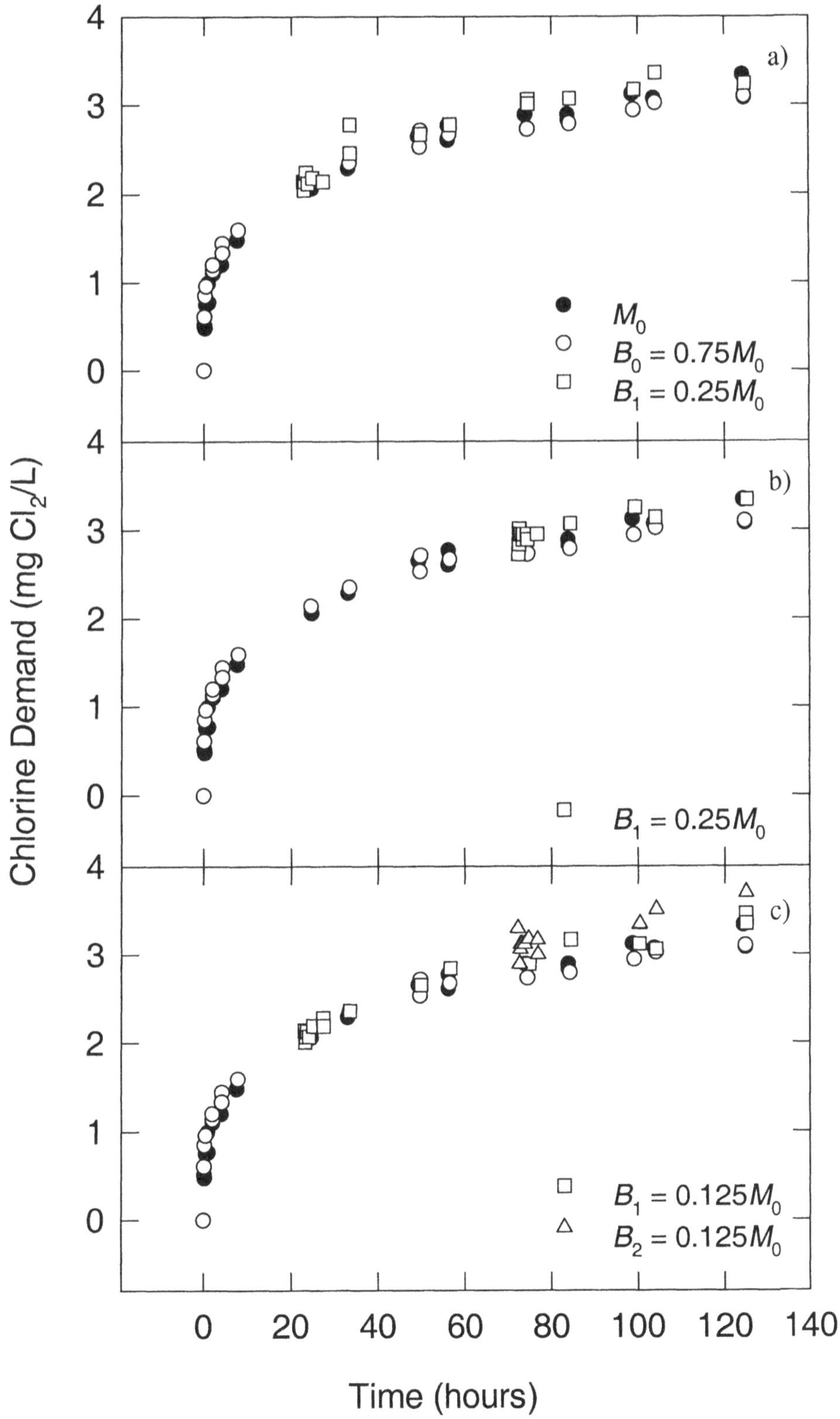

Figure C.11: Chlorine demand data of phase IIa for South Norwalk water based on a Cl:DOC of 3:1 and $B_0 = 0.75M_0$ with boost doses at a) 24 hours, b) 72 hours, and c) 24 and 72 hours.

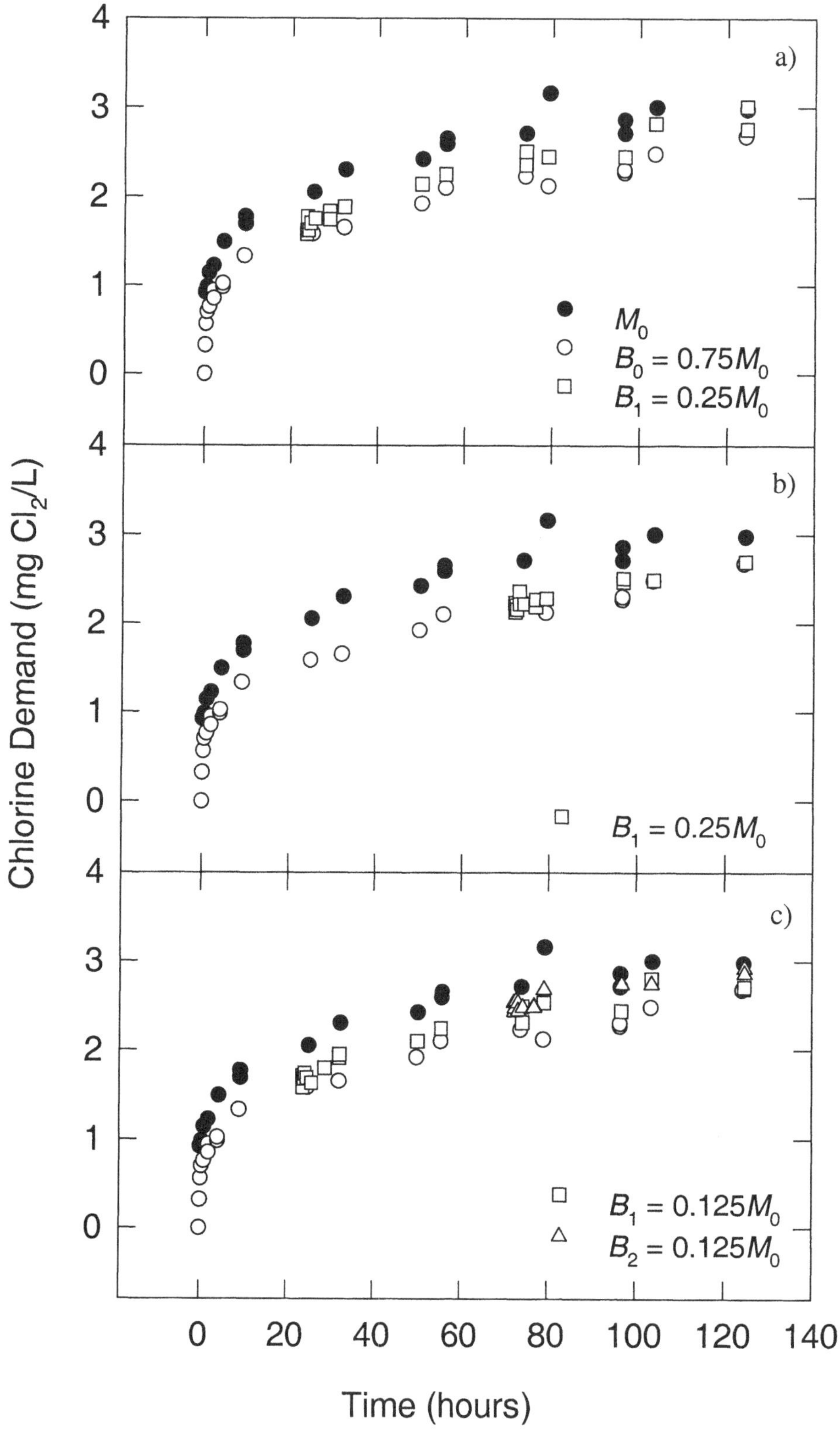

Figure C.12: Chlorine demand data of phase IIa for South Norwalk water based on a Cl:DOC of 2:1 and $B_0 = 0.75M_0$ with boost doses at a) 24 hours, b) 72 hours, and c) 24 and 72 hours.

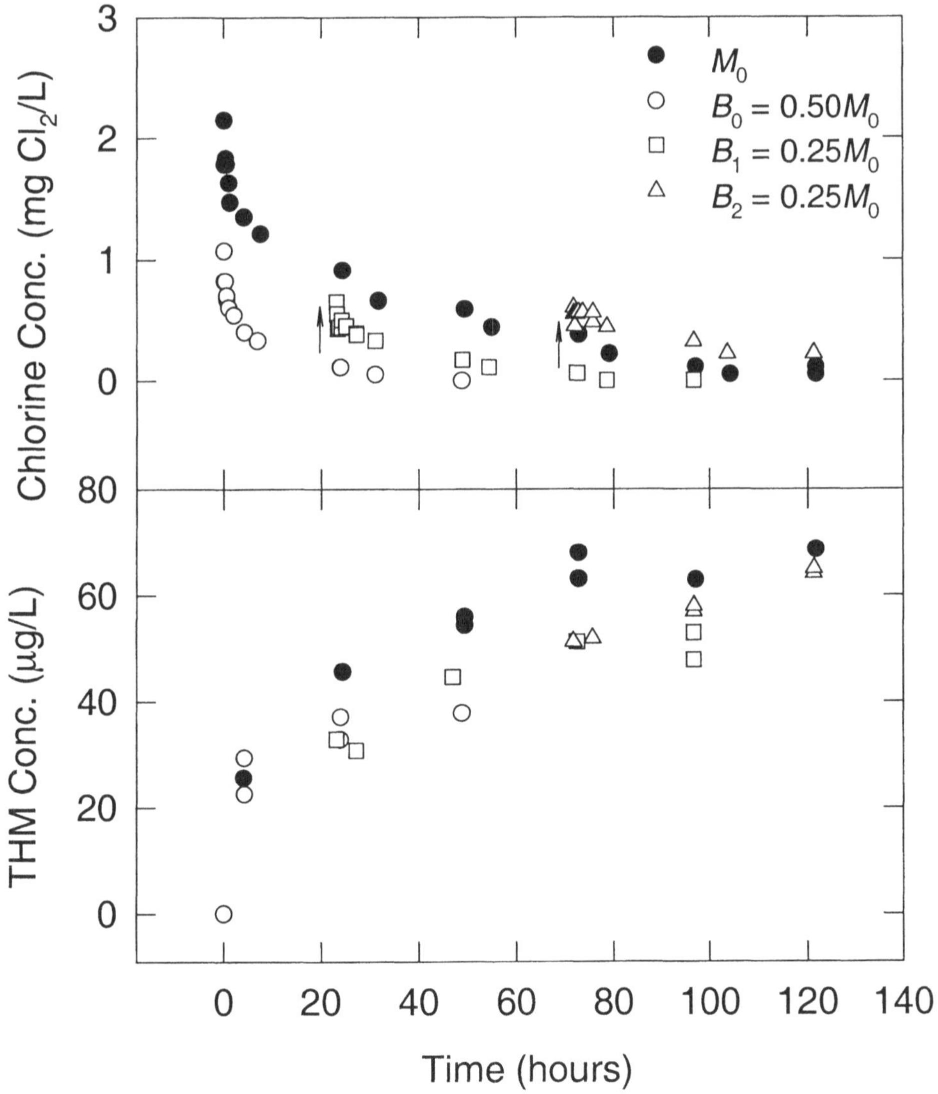

Figure C.13: Chlorine decay and THM formation data of phase IIb for South Norwalk water based on multiple boost doses.

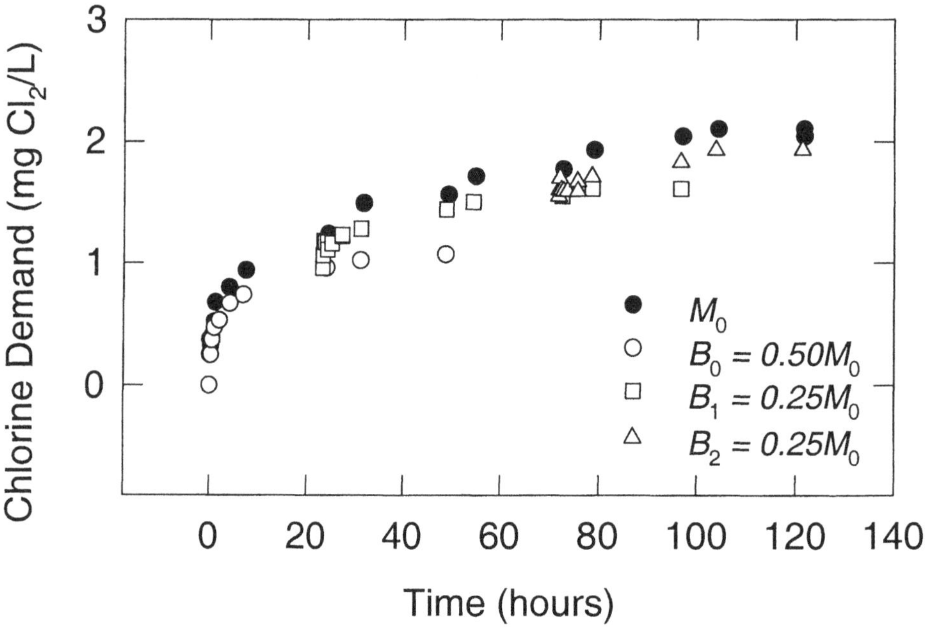

Figure C.14: Chlorine demand data of phase IIb for the South Norwalk water with multiple boost doses.

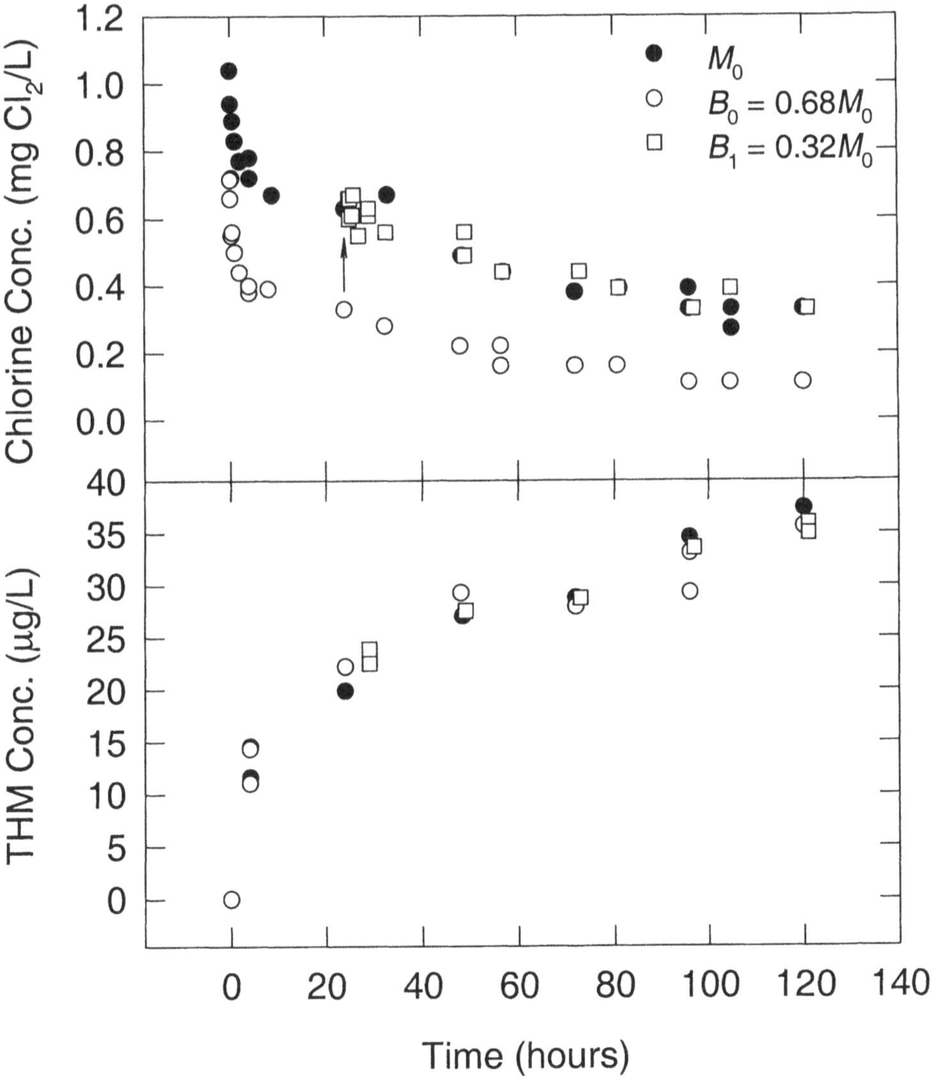

Figure C.15: Chlorine decay and THM data of phase IIb for the Cincinnati water based on $B_0 = 0.68M_0$.

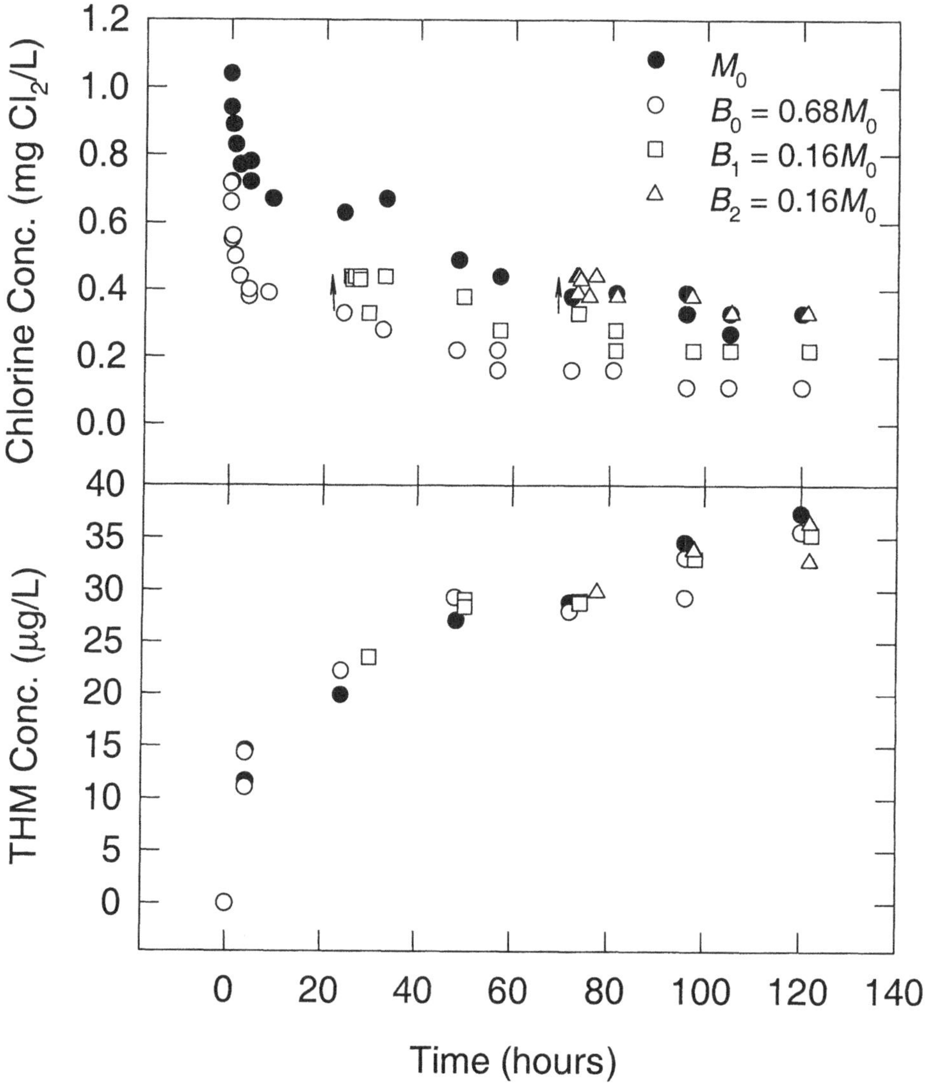

Figure C.16: Chlorine decay and THM formation data of phase IIb for Cincinnati water based on multiple boost doses.

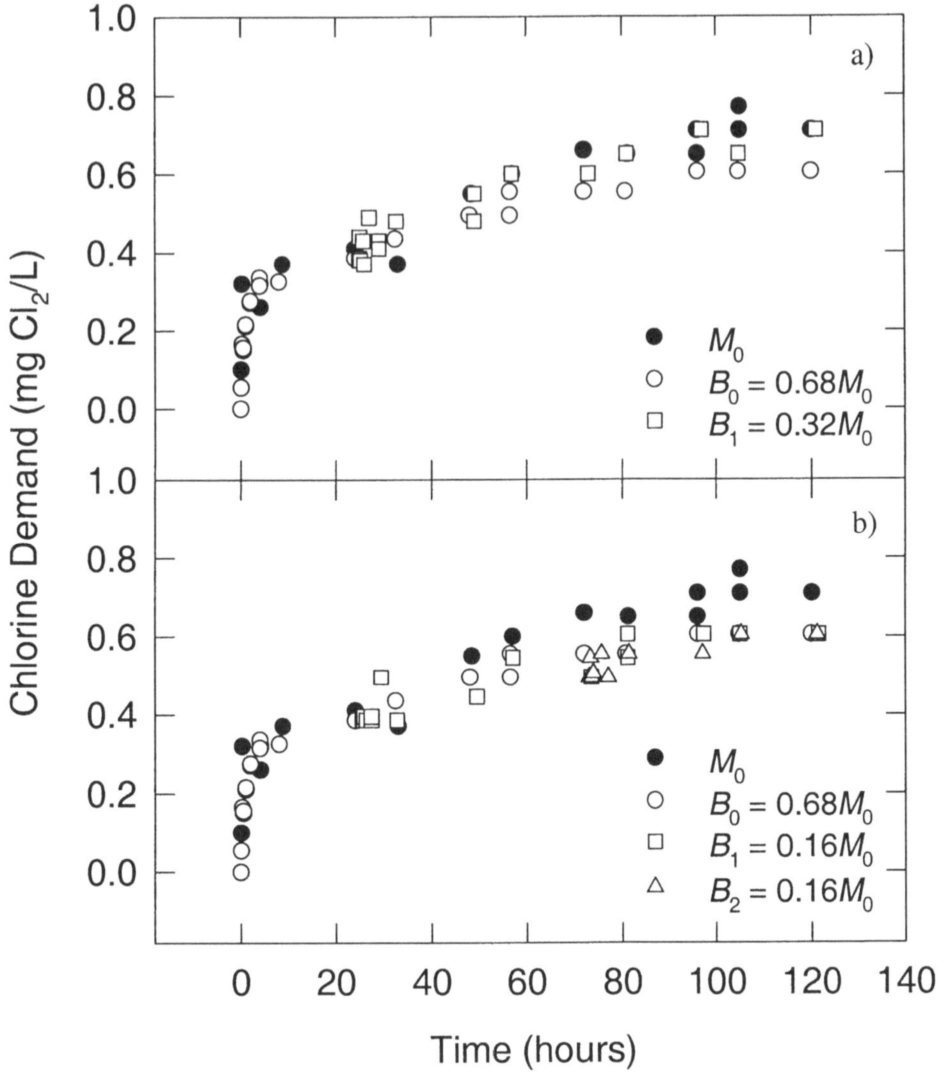

Figure C.17: Chlorine demand data of phase IIb for the Cincinnati water with a) $B_0 = 0.68M_0$ and b) multiple boost doses.

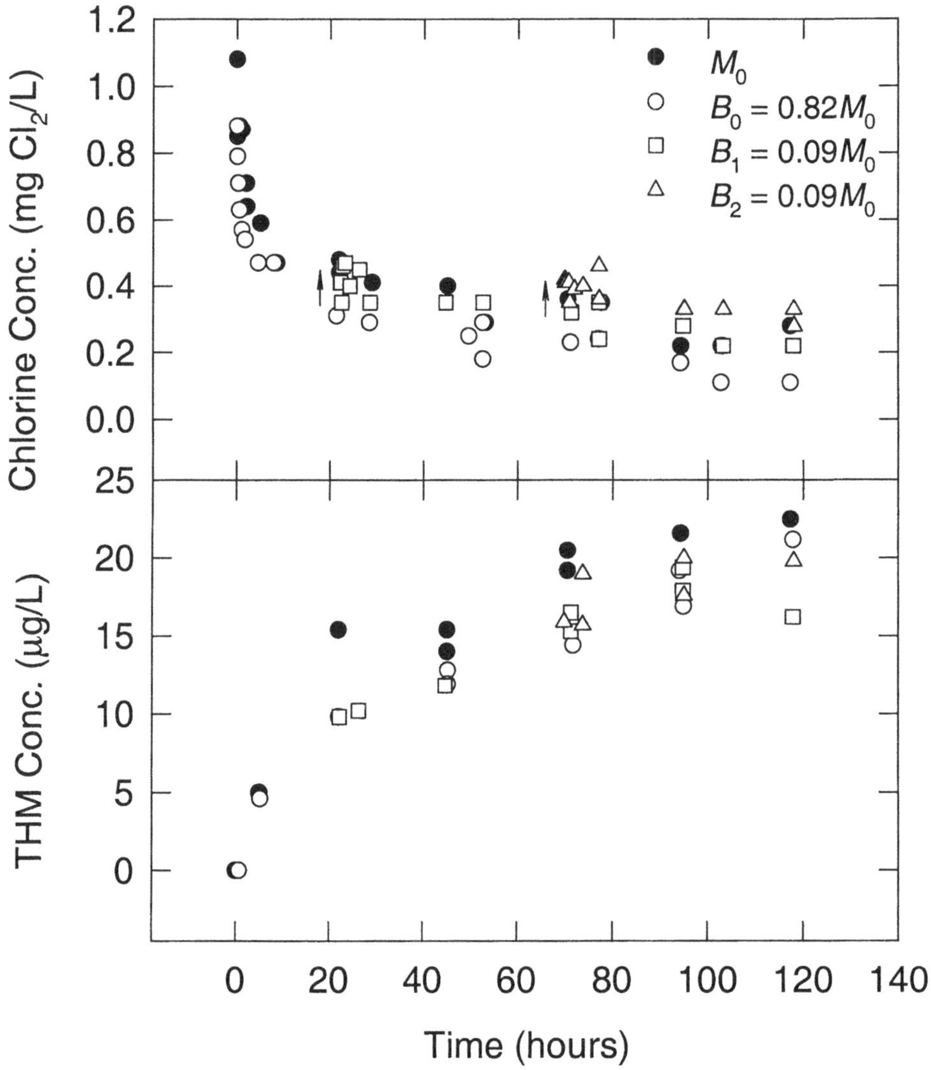

Figure C.18: Chlorine decay and THM formation data of phase IIb for Midland water based on multiple boost doses.

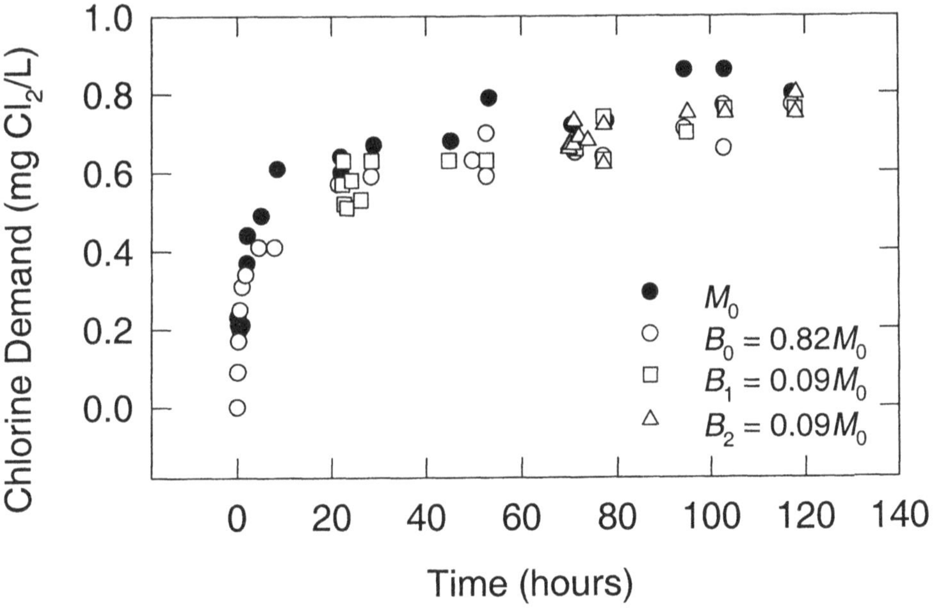

Figure C.19: Chlorine demand data of phase IIb for the Midland water with multiple boost doses.

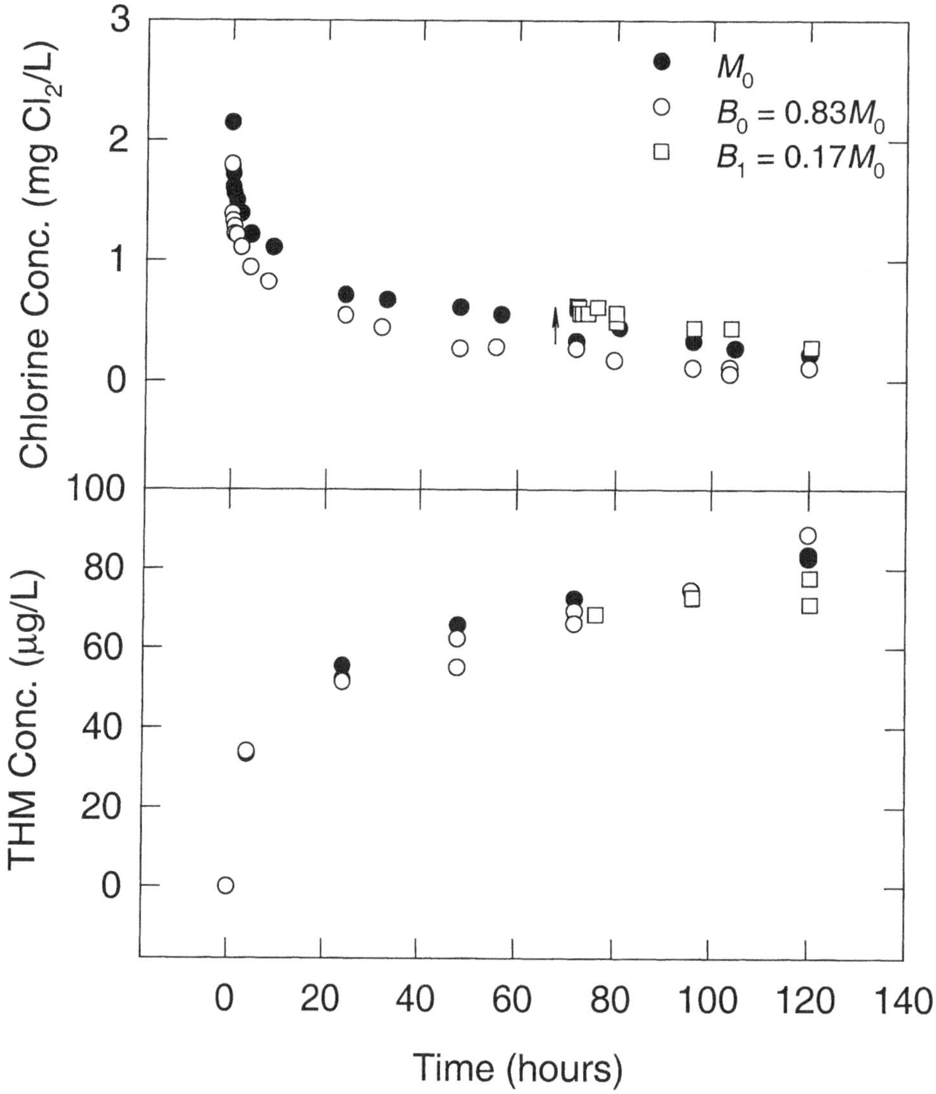

Figure C.20: Chlorine decay and THM data of phase IIb for the Las Vegas (pH = 8.0) water based on $B_0 = 0.83M_0$.

Appendix D

Linear Superposition of Booster Doses

D.1 Introduction

Given that: 1) the water distribution system hydraulics are known and time varying, 2) disinfectant decay kinetics are first-order with respect to disinfectant concentration, and 3) the reaction rate coefficients are independent of the booster injections **u**, we show that the principle of linear superposition applies to disinfectant transport in water distribution networks with time-varying disinfectant doses at the booster locations. For this purpose, it will suffice to show that network disinfectant transport can be described mathematically as a linear continuous-time dynamic system. This result is then used to derive the relationship between the cycle times of periodic network hydraulics, booster doses, and monitoring node concentrations.

D.2 Disinfectant Transport in Networks

Distribution system hydraulics (pipe flow rates and reservoir volumes) are typically described by a sequence of linked steady-state hydraulic simulations in a procedure known as extended period simulation (EPS)(Rao and Bree, 1977). Network water quality models typically idealize the distribution system as a network of plug flow reactors (PFR) without axial dispersion(Rossman and Boulos, 1996). The partial differential equation describing a PFR without axial dispersion is used to model disinfectant transport and decay in pipe p that connects nodes i and j, where flow is from node i to j:

$$\frac{\partial c_p}{\partial t} = -\frac{4q_p}{\pi D_p^2}\frac{\partial c_p}{\partial x} - bc_p, p = 1, \ldots, P \tag{D.1}$$

$$c_p(x = 0, t) = c_i^*(t) \tag{D.2}$$

$$c_p(x, t = 0) = c_p^0(x) \tag{D.3}$$

where x [L] is axial distance with positive orientation in the direction of assumed flow from node i to node j, t [T] is time, D_p [L] is the pipe diameter, q_p $[L^3/T]$ is the pipe flow rate, b [1/T] is the first-order rate constant, $c_P^0(x)$ $[M/L^3]$ is the initial pipe concentration, $c_i^*(t)$ is the concentration exiting upstream boundary node i, and P is the number of pipes in the network. These hyperbolic PFR equations can be approximated by a first-order ordinary differential system in standard state-space form, by replacing the spatial derivatives with a suitable finite-difference scheme. One such scheme (backward difference) yields:

$$\frac{dc_{p,l}}{dt} = -\frac{4q_p}{\Delta x_p \pi D_p^2}(c_{p,l} - c_{p,l-1}) - bc_{p,l}, p = 1, \ldots P, l = 1, \ldots, n_p \tag{D.4}$$

$$c_{p,0} = c_i^*(t) \tag{D.5}$$

$$c_{p,l}(t = 0) = c_{p,l}^0 \tag{D.6}$$

where $c_{p,l}$ is the species concentration at finite difference node l in pipe p, Δx_p is the uniform distance between pipe finite difference nodes, and $n_p \Delta x_p$ equals the pipe length. Pipe p has $n_p + 1$ finite difference nodes, and the boundary node index equals 0 ($x = 0$).

Complete and instantaneous mixing at each network node defines the boundary functions $c_i^*(t)$ for all pipes connected to network junctions (and not reservoirs), and provides the coupling between the PFR models for individual pipes. For an arbitrary junction node i:

$$c_i^*(t) = \frac{\sum_{p \in IN_i} q_p c_{p,n_p}(t) + u_i(t)}{\sum_{p \in IN_i} q_p} \tag{D.7}$$

where IN_i is the set of all pipe indices for which flow is directed into junction node i, and $u_i(t)$ [M/T] is the booster dose at junction node i and time t. For locations where upstream node i of pipe p is connected to the kth reservoir (and not a network junction), the reservoir concentration defines the boundary condition for pipe p,

$$c_i^*(t) = c_{r_k}(t) \tag{D.8}$$

where $c_{r_k}(t)$ is the concentration in reservoir k at time t. Reservoirs are often modeled as continuous flow stirred tank reactors (CFSTRs)(Grayman et al., 1996). (The application of linear superposition does not require that reservoirs be CFSTRs, any more than it requires that pipes be PFRs; it is only necessary that they be linear devices with respect to concentration.) Conservation of disinfectant mass for reservoir k discharging into pipe p yields:

$$\frac{dc_{r_k}}{dt} = -bc_{r_k}, \quad c_{r_k}(t = 0) = c_{r_k}^0 \tag{D.9}$$

where the disinfectant decay rate in reservoir k is assumed to be first-order. Although not affecting pipe boundary conditions, the disinfectant concentrations in filling reservoirs must be modeled because of their impact during later hydraulic periods. If the downstream node j of pipe p is connected to reservoir k, then the concentration in reservoir k is described by:

$$\frac{dc_{r_k}}{dt} = \frac{q_p}{v_k}(c_{p,n_p}(t) - c_{r_k}) - bc_{r_k}, \quad c_{r_k}(t = 0) = c_{r_k}^0 \tag{D.10}$$

where v_k is the fluid volume in reservoir k at the beginning of the current hydraulic period.

D.3 Proof of Linear Superposition

The dynamic system describing disinfectant transport and decay in the distribution network is defined by (D.4, (D.9), and (D.10) along with boundary and initial conditions, and coupling relationships (D.5–D.8). These equations are linear in pipe node and reservoir disinfectant concentrations. Defining $\mathbf{c}_j$ as the vector of all pipe node and reservoir disinfectant concentrations during the jth hydraulic period and similarly $\mathbf{u}_j$ as the vector of time-varying booster dose rates at all junction nodes, the dynamic system describing disinfectant transport within the jth hydraulic period is a nonhomogeneous linear system with constant coefficients, and can be written in standard state-space form:

$$\frac{d\mathbf{c}_j}{dt} = \mathbf{A}_j\mathbf{c}_j(t) + \mathbf{B}_j\mathbf{u}_j(t), \quad \mathbf{c}_j(0) = \mathbf{c}_{j-1}(\Delta t_h) \tag{D.11}$$

where Δt_h is the length of a hydraulic time step. The constant matrices $\mathbf{A}_j$ and $\mathbf{B}_j$ depend on the network topology and discretization, pipe flow rates, and pipe and reservoir characteristics. The general solution to (D.11) at the beginning of the $(j+1)$th hydraulic timestep can be expressed in terms of the state-transition matrix for the homogeneous system within the jth hydraulic time step $\mathbf{\Phi}_j(t,\tau)$, the initial state $\mathbf{c}_j(0)$, and the booster doses $\mathbf{u}_j(t)$ ((Luenberger, 1979), pp. 118-121),

$$\mathbf{c}_{j+1}(0) = \mathbf{\Phi}_j(\Delta t_h, 0)\mathbf{c}_j(0) + \int_0^{\Delta t_h} \mathbf{\Phi}_j(\Delta t_h, \tau)\mathbf{B}_j\mathbf{u}_j(\tau)d\tau \tag{D.12}$$

where the vector integrand is a continuous summation of the individual dose effects from $t = 0$ to Δt_h. The state-transition matrix $\mathbf{\Phi}_j(t,\tau) = \mathbf{\Psi}_j(t)\mathbf{\Psi}_j(\tau)^{-1}$, where $\mathbf{\Psi}_j$ is the fundamental matrix of the homogeneous differential equation and satisfies $d\mathbf{\Psi}_j(t)/dt = \mathbf{A}_j\mathbf{\Psi}_j(t)$, yielding $\mathbf{\Psi}_j(t) = e^{\mathbf{A}_j t}$ and thus $\mathbf{\Phi}_j(t,\tau) = e^{\mathbf{A}_j(t-\tau)}$((Luenberger, 1979), pp. 113-117). The solution (D.12) can be expressed notationally as,

$$\mathbf{c}_{j+1}(0) = \beta_j\mathbf{c}_j(0) + \sum\nolimits_j, \tag{D.13}$$

where

$$\beta_j = e^{\mathbf{A}_j\Delta t_h} \tag{D.14}$$

$$\sum\nolimits_j = \int_0^{\Delta t_h} e^{\mathbf{A}_j(\Delta t_h - t)}\mathbf{B}_j\mathbf{u}_j(\tau)d\tau \tag{D.15}$$

and $\sum_j$ represents the contribution of mass injections occurring during hydraulic period j to the concentrations at the end of that period. By writing (D.13) for sequential hydraulic periods $j = 0, \ldots, J-1$, the concentrations at the beginning of the jth period can be related to those at the beginning of the first,

$$\mathbf{c}_J(0) = \mathbf{\Omega}(J,0)\mathbf{c}_0 + \sum_{i=0}^{J-1}\mathbf{\Omega}(J,i+1)\sum\nolimits_i \tag{D.16}$$

where $\mathbf{\Omega}(k,l)$ is the state-transition matrix from the beginning of the lth hydraulic timestep to the beginning of the kth hydraulic timestep. The state-transition matrix $\mathbf{\Omega}(k,l) = \beta_{k-1}\beta_{k-2}\ldots\beta_l, k > l$, and $\mathbf{\Omega}(k,k) = \mathbf{I}$.

Since monitoring constraints are imposed only after initial condition effects have dissipated (i.e. only for times such that $J\Delta t_h \geq M\Delta t_m$), the concentrations depend only on the contributions of the mass injections, and the term $\mathbf{\Omega}(J,0)\mathbf{c}_0$ is negligible,

$$\mathbf{c}_J(0) = \sum_{i=0}^{J-1} \mathbf{\Omega}(J, i+1) \sum\nolimits_i \qquad \text{(D.17)}$$

Eq. (D.17) shows that the concentrations within any hydraulic period are expressed as a linear superposition of effects attributed to the time varying booster doses at individual junction nodes.

D.4 Concentration Periodicity

In developing the finite-time scheduling model (Chapter 5) the concentrations, and thus the impulse response coefficients, are assumed periodic over cycle time T_α. Thus a relevant question is: under what conditions are these concentrations periodic, and what factors determine their cycle time T_α? To explore this question, consider the solution (D.17) written at the beginning of hydraulic time step j which is a function of $\mathbf{A}_j$, $\mathbf{B}_j$, and $\mathbf{u}_j$. The matrices $\mathbf{A}_j$ and $\mathbf{B}_j$ depend on the network hydraulic solution in the jth period, and thus will be periodic if the hydraulic solution is periodic. Accordingly, we assume a periodic hydraulic solution with cycle time T_h, which implies a periodic cycle time of T_h for β_j and hence $\mathbf{\Omega}(k,l)$. We further assume that T_s, the scheduling cycle time for $\mathbf{u}_j$, is selected such that $\eta T_s = \mu T_h$ for some integers $\eta, \mu > 0$, from which it follows that $\sum_i$ is periodic with cycle time $\eta T_s = \mu T_h$. Eq. (D.17) is then a sum of products of periodic functions with cycle times equal to T_h or a multiple thereof; thus the concentrations, and the composite impulse response coefficients, have a cycle time $T_\alpha = \eta T_s = \mu T_h$.

Appendix E

CD-ROM Documentation

E.1 List of CD-ROM Contents

1. BDDA Version 1.00 Executable
2. BDDA Version 1.00 Source Code
3. Epanet Version 2.00 Executable for Windows
4. Epanet Toolkit Version 2.00
5. CPLEX Version 6.50 Function Library*

* A license is required to operate the function library, contact ILOG for further information (see below).

E.2 BDDA Instructions

To setup the BDDA copy the file bddav100.exe located in the BDDA Version 1.00 directory to the working directory. To run the BDDA the epanet2.dll and cplex65.dll files and the necessary data files must also be located in the working directory (alternatively, the DLL files can be placed in a system directory that is included on the DLL search path). Further instructions can be found in Appendix A.

E.3 EPANET Instructions

Epanet software for windows has been included so that the optimal solutions can be visualized using the graphical users interface. Using the "EPANET" keyword in the

[OPTIONS] section of the BDDA input file allows easy simulation of optimal booster system solutions. The BDDA writes an Epanet format input file with the optimal solution when using the commercial CPLEX solver; otherwise cross reference information is written so that the optimal solution provided by another solver can be interpreted and the simulation set up by hand.

The BDDA software is an Epanet Toolkit application (Rossman 1999). The toolkit has been included to facilitate future modifications to the BDDA software.

E.4 CPLEX Instructions

A license is required to run CPLEX. ILOG may be contacted to arrange licensing agreements at:

ILOG Contact Center (USA and Canada)
Web: http://www.ilog.com
E-mail: info@ilog.com
Toll-free: 800-367-4564
Tel: 775-881-2800

E.5 Public Domain LP and MILP Solvers

The public domain LP and MILP solvers "lp_solve" and "MINOS" are available on the web at: http://www.netlib.org

These solvers may perform well for certain small to medium sized problems, but the commercial CPLEX software will likely be required for serious applications.

This study was funded by the Awwa Research Foundation (AwwaRF). AwwaRF assumes no responsibility for the content of this CD-ROM or for the opinions or statements of fact expressed on this CD-ROM. The mention of trade names for commercial products does not represent or imply the approval or endorsement of AwwaRF. This CD-ROM is presented solely for informational purposes.

Bibliography

Adin, A., Katzhendler, J., Alkaslassy, D., and Rav-Acha, C. (1991). Trihalomethane formation in chlorinated drinking water. a kinetic model. *Water Res.*, 25(7):797–805.

Agency, U. S. E. P. (1990). *Guidance Manual for Compliance with the Filtration and Disinfection Requirements for Public Water Systems Using Surface Waters.* ISBN 0-89867-558-8.

Amy, G. L., Chadik, P. A., and Chowdhury, Z. K. (1987). Developing models for predicting trihalomethane formation potential and kinetics. *J. Amer. Water Works Assoc.*, 79(7):89–97.

APHA, AWWA, and WEF (1995). *Standard Methods for the Examination of Water and Wastewater.* American Public Health Assoc., Washington, DC., 19th ed. edition. A. D. Eaton, L. S. Clesceri, and A. E. Greenberg, eds.

Auerbach, J. (1994). Costs and benefits of current SDWA regulations. *Journal of the American Water Works Association*, 86(2):69–78.

AWWA (1992a). Survey of water utility disinfection practices. *Journal of the American Water Works Association*, 84(9):121–128.

AWWA (1992b). Survey of water utility disinfection practices. *J. Amer. Water Works Assoc.*, 84(9):121–128.

Bielmeier, S. R., Best, D. S., Guidici, D. L., and Narotsky, M. G. (2001). Pregnancy loss in the rat caused by bromodichloromethane. *Toxicological Sciences*, 59(2):309–315.

Biswas, P., Lu, C., and Clark, R. M. (1993). A model for chlorine concentration decay in pipes. *Water Res.*, 27(12):1715–1724.

Boccelli, D., Tryby, M., Koechling, M., Uber, J., Summers, R. S., and Rossman, L. (1997). *Proceedings of the AWWA Annual Conference and Exposition*, chapter Bulk Decay Kinetics of Rechlorinated Water. American Water Works Association, Denver, CO.

Boccelli, D., Tryby, M., Uber, J., Rossman, L., Zierolf, M., and Polycarpou, M. (1998). Optimal scheduling of booster disinfection in water distribution systems. *ASCE Journal of Water Resources Planning and Management*, 124(2):99–110.

Boccelli, D. L. (1999). Booster disinfection: Bulk decay kinetics, by-product formation, and model development. Master's thesis, University of Cincinnati, Cincinnati, Ohio.

Brereton, J. A. and Mavinic, D. S. (1997). *Proceedings of the A.W.W.A. Annual Conference*, chapter The 'Material Specific' simulated distribution system test for disinfection by-product formation potential, pages 699–706. Amer. Water Works Assoc.

Brereton, J. A., Mavinic, D. S., and Crowe, B. (1997). *Proceedings of the A.S.C.E. National Conference on Environmental Engineering*, chapter Chlorine demand of tuberculated cast iron: impact on trihalomethane formation within distribution systems, pages 667–668. Amer. Soc. of Civil Engineers.

Bryant, E. A., Fulton, G. P., and Budd, G. C. (1992). *Disinfection Alternatives for Safe Drinking Water.* Van Nostrand Reinhold, New York, NY.

Bull, R. J. and Kopfler, R. C. (1991). *Health Effects of Disinfectants and Disinfection By-Products.* American Water Works Association Research Foundation.

Chowdhury, Z. K., Amy, G. L., and Siddiqui, M. (1991). Modeling effects of bromide ion concentration on the formation of brominated trihalomethanes. In *Proceedings of the Annual Conf., Amer. Water Works Assoc.*, Philadelphia, PA.

Chowdhury, Z. K., Dickerson, L., and Amy, G. L. (1993). Modeling the formation of haloacetic acids and trihalomethanes in chemically coagulated waters. In *Proceeding of the Water Quality Tech. Conf., Amer. Water Works Assoc.*, Miami, FL.

Church, R. and Revelle, C. (1974). The maximal covering location problem. *Papers of the Regional Science Association*, 32:101–118.

Clark, R., Adams, J., and Lykins, B. (1994). Dbp control in drinking water: Cost and performance. *A.S.C.E. Journal of Environmental Engineering*, 120(4):759–782.

Clark, R., Grayman, W., Males, R., and Hess, A. (1993). Modeling contaminant propagation in drinking water distribution systems. *ASCE Journal of Environmental Engineering*, 119(2):349–364.

Clark, R. M. Chlorine demand and TTHM formation kinetics: A second-order model. *A.S.C.E. J. of Environmental Eng.*, 124(1).

Clark, R. M. and Sivaganesan, M. (1998). Predicting chlorine residuals and formation of TTHMs in drinking water. *A.S.C.E. J. of Env. Eng.*, 124(12):1203–1210.

Cohon, J. L. (1978). *Multiobjective programming and planning.* academic press, New York, NY.

Cowman, G. A. and Singer, P. C. (1996). Effect of bromide ion on haloacetic acid speciation resulting from chlorination and chloramination of aquatic humic substances. *Environmental Science and Technology*, 30(1):16–24.

Diehl, A. C., Speitel, G. E., Symons, J. M., Krasner, S. W., and Hwang, C. J. (1995). Factors affecting disinfection by-product formation during chloramination. In *Proceedings of the 1995 A.W.W.A. Annual conference*, American Water Works Assoc., Denver, CO.

Dugan, N. R., Summers, R. S., Miltner, R. J., and Shukairy, H. M. (1995). An alternative approach to predicting chlorine residual decay. In *Proceeding of the Water Quality Tech. Conf., Amer. Water Works Assoc.*, New Orleans, LA.

Edge, J. C. and Finch, P. E. (1987). Observations on bacterial aftergrowth in water supply distribution systems. *J. Institution of Water and Env. Manag.*, 1(1):104–110.

Edzwald, J. K., Becker, W. C., and Wattier, K. L. (1985). Surrogate parameters for monitoring organic matter and THM precursors. *J. Amer. Water Works Assoc.*, 77(4):122–132.

Efroymson, M. and Ray, T. (1968). A branch-and-bound algorithm for plant location. *Operations Research*, 14.

El-Shafy, M. A. and Grünwald, A. (2000). THM formation in water supply in South Bohemia, Czech Republic. *Water Res.*, 34(13):3453–3459.

Fleischacker, S. J. and Randtke, J. (1983). Formation of organic chlorine in public water supplies. *J. American Water Works Assoc.*, 75(3):132–138.

Geldreich, E. E. (1996). *Microbial Quality of Water Supply in Distribution Systems.* Lewis Publishers, New York, NY.

Gibbs, R. A., Scutt, J. E., and Croll, B. T. (1990). Microbiological and trihalomethane responses to booster chlorination. *J. Institution of Water and Env. Manag.*, 4(4):131–139.

Grayman, W. M., Deininger, R. A., Green, A., Boulos, P. F., Bowcock, R. W., and Godwin, C. C. (1996). Water quality and mixing models for tanks and reservoirs. *Journal of the American Water Works Association*, 88(7):60–73.

Haas, C. N. (1999). Benefits of using a disinfectant residual. *Journal of the American Water Works Association*, 91(1):65–69.

Haas, C. N. and Karra, S. M. (1984). Kinetics of wastewater chlorine demand exertion. *J. Water Pollution Control Fed.*, 56(2):170–173.

Harrington, G. W., Chowdhury, Z. K., and Owen, D. M. (1992). Developing a computer model to simulate DBP formation during water treatment. *J. Amer. Water Works Assoc.*, 84(11):78–87.

Ireland, C. and Knudson, M. (1998). *Protecting Water Quality in the Distribution System: What is the Role of Disinfection Residuals?*, chapter Portland's Experience with Chloramine Residual Management. Amer. Water Works Assoc. and Intl. Water Supply Assoc. Proceedings of Symposium held in Philadelphia, PA.

Jadas-Hecart, A., Morer, A. E., Stitou, M., Bouillot, P., and Legube, B. (1992). Modelisation de la demande en chlore d'une eau traitee. *Water Research*, 26(8):1073–1084.

Jafvert, C. T. and Valentine, R. L. (1992). Reaction scheme for the chlorination of ammoniacal water. *Environmental Science and Technology*, 26(3):577–585.

Kiene, L., Levi, Y., Lu, W., and Moncet, C. (1998). *Protecting Water Quality in the Distribution System: What is the Role of Disinfection Residuals?*, chapter Combining Optimization Models, On-Line Sensors and Booster Chlorination to Manage Water Quality in Distribution Systems. Amer. Water Works Assoc. and Intl. Water Supply Assoc. Proceedings of Symposium held in Philadelphia, PA.

Koechling, M. T., Summers, R. S., Dugan, N. R., Shukairy, H. M., and Miltner, R. J. (1997). Modeling chlorine decay using ultraviolet absorbance. In *Proceedings of the Annual Conf., Water Research, Amer. Water Works Assoc.*, Atlanta, GA.

Krasner, S. W., McGuire, M. J., Jacangelo, J., Patania, N., and Reagan, K. (1989). The occurrence of disinfection by-products in US drinking water. *Journal of the American Water Works Assoc.*, 81(8):41–53.

Krasner, S. W., Sclimenti, M. J., Chinn, R., Chowdhury, Z. K., and Owen, D. M. (1996). *Disinfection By-Products in Water Treatment: The Chemistry of their Formation and Control*, chapter The impact of TOC and Bromide on chlorination by-product formation. Lewis Publishers, New York, NY. R. A. Minear and G. L. Amy, Eds.

Larson, R. A. and Weber, E. J. (1994). *Reaction Mechanisms in Environmental Organic Chemistry*, chapter Reactions with Disinfectants, pages 275–357. Lewis Publishers, Boca Raton, FL.

LeChevallier, M. W., Babcock, T. M., and Lee, R. G. (1987). Examination and characterization of distribution system biofilms. *Applied and Environmental Microbiology*, 53(12):2714–2724.

Levenspiel, O. (1972). *Chemical Reaction Engineering*. John Wiley and Sons, New York, NY, 2nd edition edition.

Levenspiel, O. (1979). *Michaelis-Menton Kinetics.* The chemical reactor omnibook. OSU Book stores, Corvallis, OR.

Lieu, N. I., Wolfe, R. L., and Means, E. G. (1993). Optimizing chloramine disinfection for the control of nitrification. *Journal of the American Water Works Association*, 85(2):84–90.

Luenberger, D. G. (1979). *Introduction to Dynamic Systems.* John Wiley and Sons, New York, NY.

Lyn, T. L. and Taylor, J. S. (1993). Modeling compliance of chlorine residual and disinfection by-products. In *Proceedings of the water quality tech. conf., Amer. Water Works Assoc.*, Miami, FL.

Mann, R. W., Wolfe, R. L., Barrett, S. E., and Lieu, N. I. (1998). Optimizing chloramine residuals to prevent nitrification in southern california water utilities. In *Protecting Water Quality in the Distribution System: What is the Role of Disinfection Residuals?*, American Water Works Assoc., Denver, CO.

Matheis, W., Koechling, M. T., Summers, R. S., and Frimmel, F. H. (1998). *Proceedings of the AWWA Annual Conference*, chapter Influence of drinking water treatment on chlorine decay kinetics, pages 539–564. Amer. Water Works Assoc.

McGuire, M. J., Davis, M. K., Tate, C. H., Aieta, E. M., Howe, E. W., and Crittenden, J. C. (1991). Evaluating gac for trihalomethane control. *J. Amer. Water Works Assoc.*, 83(1):38–48.

Meyer, P. and E. D. Brill, J. (1988). A method for locating wells in a groundwater monitoring network under conditions of uncertainty. *Water Resource Research*, 24(8):1277–1282.

Miltner, R. J., Summers, R. S., Dugan, N. R., Koechling, M. T., and Moll, D. M. (1996). *Proceedings of the Water Quality Technology Conference*, chapter A comparative evaluation of biological filters. Amer. Water Works Assoc.

Morris, J. C. and Baum, B. (1978). *Water Chlorination: Environmental Impact and Health Effects*, volume Vol. 2, chapter Precursors and Mechanisms: Haloform Formation in the Chlorination of Water Supplies, pages 29–48. Ann Arbor Science Pub., Ann Arbor, MI.

Morrison, R. T. and Boyd, R. N. (1987). *Organic Chemistry.* Allyn and Bacon, Inc., Boston, MA, 5th ed. edition.

Neden, D. G., Jones, R. J., Smith, J. R., Kirmeyer, G. J., and Foust, G. W. (1992). Comparing chlorination and chloramination for controlling bacterial regrowth. *Journal of the American Water Works Association*, 84(7):80–88.

Nemhauser, G. L. and Wolsey, L. S. (1988). *Integer and Combinatorial Optimization.* John Wiley and Sons, New York, NY.

Nguyen, B. and Montiel, A. (1994a). On line quality control in distribution networks. *Water Supply*, 12(3/4):289–299.

Nguyen, B. and Montiel, A. (1994b). Online quality control in distribution networks. *Water Supply*, 12(2):289–299.

Norman, T. S., Harms, L. L., and Looyenga, R. W. (1980). The use of chloramines to prevent trihalomethane formation. *Journal of the American Water Works Association*, 72(3):176–180.

Owen, D. M., Chowdhury, Z. K., and Krasner, S. W. (1992). Predicting THM formation in bromide-rich waters. In *Proceedings of the water quality tech. conf., Amer. Water Works Assoc.*

Powell, J. C., West, J. R., Hallam, N. B., Forster, C. F., and Simms, J. (2000). Performance of various kinetic models for chlorine decay. *A.S.C.E. J. Wat. Res. Plan. Mngmt*, 126(1):13–20.

Rao, H. S. and Bree, D. W. (1977). Extended period simulation of water systems - part a. *Journal of The Hydraulics Divisions*, 103:97–108.

Rittman, B. E. and Snoeying, V. L. (1984). Achieving biologically stable drinking water. *Journal of the American Water Works Association*, 76(10):106–114.

Rossman, L. and Boulos, P. (1996). Numerical methods for modeling water quality in distribution systems: A comparison. *Journal of Water Resources Planning and Management*, 122(2):137–146.

Rossman, L., Clark, R., and Grayman, W. (1994). Modeling chlorine residuals in drinking-water distribution systems. *ASCE Journal of Environmental Engineering*, 120(4):803–820.

Rossman, L. A. (1999). The epanet programmer's toolkit. In *Proceedings of Water Resources Planning and Management Division Annual Specialty Conference, ASCE*, Tempe, AZ.

Rossman, L. A. (2000). *EPANET User's Manual*. Risk Reduction Engineering Laboratory, U.S. Environmental Protection Agency, Cincinnati, Ohio.

Shaw, E. S. and Regli, S. (1999). U.s. regulations of residual disinfection. *Journal of the American Water Works Association*, 91(1):75–80.

Shorney, H. L. and Knocke, W. R. (1996). *Bench-scale evaluation of manganese stability in the presence of chlorine and combined chlorine*. Amer. Water Works Assoc., Boston, MA. Poster Presentation, Water Quality Technology Conference.

Singer, P. (1994). Control of disinfection by-products in drinking water. *A.S.C.E. Journal of Environmental Engineering*, 120(4):727–744.

Singer, P. C., Barry, J. J., Palen, G. M., and Scrivner, A. E. (1981). Trihalomethane formation in North Carolina drinking waters. *J. Amer. Water Works Assoc.*, 73(8):392–401.

Singer, P. C. and Chang, D. S. (1989). Correlations between trihalomethanes and total organic halides formed during water treatment. *J. Amer. Water Works Assoc.*, 81(8):61–65.

Sinha, S., Amy, G. L., and Sohn, J. (1997). *Proceedings of the AWWA Annual Conference*, chapter Reactivity of NOM fractions in forming chlorinated DBPs, pages 329–353. Amer. Water Works Association, Denver, CO.

Snoeyink, V. L. and Jenkins, D. (1980). *Water Chemistry.* John Wiley and Sons, New York, NY.

Sohn, J., Amy, G., Boireau, A., Doudin, G., Gatel, D., and Cavard, J. (1997). *Proceedings of the Water Quality Technology Conference*, chapter In-Plant versus Distribution-System Formation of Chlorination By-Products. Amer. Water Works Assoc.

Stevens, A. A., Moore, L. A., and Miltner, R. J. (1989). Formation and control of non-trihalomethane disinfection by-products. *Journal of the American Water Works Association*, 81(8):54–60.

Summers, R. S., Hooper, S. M., Shukairy, H. M., Solarik, G., and Owen, D. (1996). Assessing DBP yield: Uniform formation conditions. *J. Amer. Water Works Assoc.*, 88(6):80–93.

Symons, J. M., Fu, P. L. K., Dressman, R. C., and Stevens, A. A. (1987). The influence of bromide ion on organic chlorine and organic bromine formation during free chlorination. *J. Amer. Water Works Assoc.*, 79(9):114–118.

Symons, J. M., Krasner, S. W., Sclimenti, M. J., Simms, L. A., and Sorenson, H. W. (1996a). *Disinfection By-Products in Water Treatement: The Chemistry of their Formation and Control*, chapter Influence of bromide ion on trihalomethane and haloacetic acid formation. Lewis Publishers, New York, NY.

Symons, J. M., Speitel, G. E., Diehl, A. C., and H. W. Sorenson, J. (1994). Precursor control in waters containing bromide. *J. Amer. Water Works Assoc.*, 86(6):48–60.

Symons, J. M., Xia, R., Diehl, A. C., Speitel, G. E., Hwang, C. J., Krasner, S. W., and Barrett, S. E. (1996b). *Water Disinfection and Natural Organic Matter: Characterization and Control*, chapter The influence of operational variables on the formation of dissolved organic halogen during chloramination. American Chemical Society, Washington, DC.

Trussell, R. R. and Umphres, M. D. (1978). The formation of trihalomethanes. *J. Amer. Water Works Assoc.*, 70(11):604–612.

Tryby, M., Boccelli, D., Koechling, M., Uber, J., Summers, R. S., and Rossman, L. (1999). Booster chlorination for managing disinfectant residuals. *J. of the Amer. Water Works Assoc.*, 91(1):95–108.

USEPA (1992). Water treatment plant simulation program user's manual version 1.21. Technical report, U.S.E.P.A., Drinking Water Technology Branch, Drinking Water Standards Division, Office of Ground Water and Drinking Water. prepared by Malcolm Pirnie, Inc.

Valentine, R. L., Ozekin, K., and Vikesland, P. J. (1997). *Chloramine Decomposition in Distribution System and Model Waters.* American Water Works Association Research Foundation, Denver, CO.

van der Kooij, D. J., Hein, J., van Lieverloo, M., Schellart, J., and Hiemstra, P. (1999). Maintaining quality without a disinfectant residual. *Journal of the American Water Works Association*, 91(5):55–64.

Vasconcelos, J. J., Boulos, P. F., Grayman, W. M., Kiene, L., Wable, O., Biswas, P., Bhari, A., Rossman, L. A., Clark, R. M., and Goodrich, J. A. (1996). Characterization and modeling of chlorine decay in distribution systems. Denver, co, AWWA Research Foundation.

Wable, O., Dumoutier, N., Duguet, J. P., Jarrige, P. A., Gelas, G., and Depierre, J. F. (1991). Modeling chlorine concentrations in a network and applications to paris distribution network. In *Proceedings of the Water Quality Modeling in Distribution Systems Symposium, Amer. Water Works Assoc.*, Cincinnati, OH.

Waller, K., Swan, S. H., DeLorenze, G., and Hopkins, B. (1998). Trihalomethanes in drinking water and spontaneous abortion. *Epidemiology*, 9(2):134–140.

White, G. C. (1992). *Handbook of Chlorination and Alternative Disinfectants.* Van Nostrand Reinhold, New York, NY, 3rd ed. edition.

Wolfe, R. L., Means, E. G., Davis, M. K., and Barrett, S. E. (1988). Biological nitrification in covered reservoirs containing chloraminated water. *Journal of the American Water Works Assocation*, 80(9):109–114.

Zierolf, M., Polycarpou, M., and Uber, J. (1998). Development and auto-calibration of an input-output model of chlorine transport in drinking water distribution systems. *IEEE Trans. on Control Systems Technology*, 6(4):543–553.